PENGUIN BOOKS

# THE FELLOWSHIP

John Gribbin is one of today's greatest writers of popular science and the author of bestselling books including *In Search of Schrödinger's Cat*, *Stardust*, *Science: A History* and *Deep Simplicity*. He is famous to his many fans for making complex ideas simple, and says that his aim in his writing – much of it done with his wife Mary Gribbin – is to share with his readers his sense of wonder at the strangeness of the univese. John Gribbin trained as an astrophysicist at Cambridge University and is currently Visiting Fellow in Astronomy at the University of Sussex. He also enjoys working on science-fiction stories in his spare time, and does most of his writing in a shed in his back garden.

JOHN GRIBBIN

# The Fellowship

## *The Story of a Revolution*

PENGUIN BOOKS

PENGUIN BOOKS

Published by the Penguin Group
Penguin Books Ltd, 80 Strand, London WC2R 0RL, England
Penguin Group (USA) Inc., 375 Hudson Street, New York, New York 10014, USA
Penguin Group (Canada), 90 Eglinton Avenue East, Suite 700, Toronto, Ontario, Canada M4P 2Y3
(a division of Pearson Penguin Canada Inc.)
Penguin Ireland, 25 St Stephen's Green, Dublin 2, Ireland
(a division of Penguin Books Ltd)
Penguin Group (Australia), 250 Camberwell Road, Camberwell, Victoria 3124, Australia
(a division of Pearson Australia Group Pty Ltd)
Penguin Books India Pvt Ltd, 11 Community Centre, Panchsheel Park, New Delhi – 110 017, India
Penguin Group (NZ), cnr Airborne and Rosedale Roads, Albany, Auckland 1310, New Zealand
(a division of Pearson New Zealand Ltd)
Penguin Books (South Africa) (Pty) Ltd, 24 Sturdee Avenue, Rosebank, Johannesburg 2196, South Africa

Penguin Books Ltd, Registered Offices: 80 Strand, London WC2R 0RL, England

www.penguin.com

First published by Allen Lane 2005
Published in Penguin Books 2006
1



Typeset by Rowland Phototypesetting Ltd, Bury St Edmunds, Suffolk
Printed in England by Clays Ltd, St Ives plc



ISBN-13: 978-0-141-01570-5
ISBN-10: 0-141-01570-5

*Men are deplorably ignorant with respect to natural things, and modern philosophers, as though dreaming in the darkness, must be aroused and taught the uses of things, the dealing with things; they must be made to quit the sort of learning that comes only from books, and that rests only on vain arguments from probability and upon conjectures . . .*

*William Gilbert*
De Magnete, *1600*

# *Contents*

# *List of Illustrations*

Photographic acknowledgements are given in parentheses

# *Acknowledgements*

As ever, Mary Gribbin played a significant role in the researching and writing of this book. The authorial 'we' in the text refers to both of us, and sometimes to the reader as well; the occasional use of 'I' indicates the personal view of JG. We are grateful to the Alfred C. Munger Foundation for a grant towards our research and travel expenses, and to the following institutions for providing access to their records: Bodleian Library, Oxford; British Library, London; King's College, Cambridge; Observatoire de Paris; Public Record Office, Kew; Royal Observatory, Greenwich; Royal Astronomical Society; Royal Society; Trinity College, Cambridge; University of Cambridge Library.

The University of Sussex continued to provide us with a base from which to work.

# *Introduction*

***Wednesday, 28 November 1660***

*It is just six months since Charles II landed at Dover en route to London, summoned by Parliament to take the crown which had once adorned the head of his father. One side effect of the restoration of the monarchy has been to bring together two groups of men whose lives have been kept apart by the events of the Civil War and the Parliamentary Interregnum, but who share a common interest in finding out how the world works. One group, headed by John Wilkins, has been based in Oxford during the Rule of Parliament; but many of them have now lost their jobs to Royalist sympathizers and gravitated to London seeking other opportunities. The other group are, by and large, Royalists who have had interesting lives during the King's exile, but have, naturally, not been at the centre of things in England; they are now eager to take up the new opportunities provided for them in London by the return of the King.*

*The place where these two groups of men have come together is Gresham College, founded by Sir Thomas Gresham in 1596. By 1660, this was the leading seat of learning in London, with professors of law, medicine, rhetoric, music, chemistry, and astronomy who gave lectures that were open to the public (or at least, the gentlemanly public of polite society). It was natural that gentlemen who were interested in what we would now call science, whatever their background, would attend these lectures, and get to know their kindred spirits. Some of these gentlemen took to meeting up after the lectures, enjoying the newly fashionable taste of coffee and discussing what they had just heard. Over the past few weeks, an idea has been brewing, carefully fostered by Wilkins, the brother-in-law of Oliver*

*Cromwell and formerly Warden of Wadham College, in Oxford. Instead of just talking about science (or natural philosophy, as they called it), why not do something about it? Why not form themselves into a society which would promote the use of experiments to probe nature and unlock her secrets?*

*Individual scientists, as we shall see, had already realized the importance of the experimental method in science, and had achieved isolated successes through its application. But this was a bold new proposal for a concerted attack, on a broad front, to find out how the world worked. In making this proposal, the Gresham group were consciously following the teaching (but not the practice) of the philosopher Francis Bacon, who was no experimental scientist himself but had written influential books promoting the idea of experimental science earlier in the seventeenth century. So it is at this meeting, on 28 November 1660, in rooms at Gresham College following a lecture on astronomy by Christopher Wren, that this particular group of natural philosophers have arranged to meet under the chairmanship of John Wilkins. Their intention is to formally constitute themselves as a society, with official minutes kept from the outset that record their objective that 'they might doe something answerable here for the promoting of Experimentall Philosophy'. There are just a dozen men present to put their names to this modest ambition, but the society they found will become the catalyst for the scientific revolution; a revolution that happens because they are the right people, in the right place, at the right time . . .*

## THE REASON WHY

The seventeenth century in Britain was one of unparalleled scientific discovery. Why? Why the seventeenth century, and why Britain? The timing is straightforward to explain as part of the Renaissance, and although an explanation of the timing of the Renaissance and the reasons for this revival of culture in Western Europe lies outside the scope of the present book, as I have argued in *Science: A History* a convenient marker for the start of the scientific revolution that would transform first Europe and then the rest of the world is 1543, the year in which Andreas Vesalius published *De Humani Corporis Fabrica* (*On the Structure of the Human Body*) and Nicolaus Copernicus published *De Revolutionibus Orbium Coelestium* (*On the Revolutions of Heavenly Bodies*). Copernicus was Polish, but had studied in Italy; Vesalius was from Brussels, studied briefly in Paris, and carried out his greatest work in Italy. Italy was the birthplace of the Renaissance, which was fuelled by both scholars and manuscripts making the short journey from Constantinople around the time of the fall of Byzantium, in the middle of the fifteenth century. Other key (and related) developments brought about what has been called the First Industrial Revolution (I would prefer 'technological' rather than 'industrial') with the introduction in Europe of moveable type and the printing press, gunpowder, and the magnetic compass. These changed the intellectual environment both by improving communications and providing information about new and exciting places, and by showing that the application of science could have practical benefits.

Fifty years after the publication of the books by Vesalius and Copernicus, Italy was still the centre of the scientific world, and in the early seventeenth century Galileo famously set out the stall of what we now call science, spelling out the key importance of experiment and observation, testing hypotheses rather than simply arguing in philosophical or logical terms about which idea is more elegant as an explanation of, say, the flight of a cannonball through the air or the passage of Mars across the night sky. In fact, Galileo was not the first person to state these ideas explicitly, and he would not have called himself a scientist, nor used the word 'science', which wasn't coined until the nineteenth century. It stems from the Latin *scientia*, meaning 'knowledge'; a term which Galileo and his contemporaries would have been thoroughly familiar with, although in a much broader context than *our* science. They would have regarded themselves as natural philosophers, the heirs to Greeks such as Pythagoras and Aristotle. But as the old saying goes, if it looks like a duck, flies like a duck, and quacks like a duck, then it *is* a duck. In modern terminology, the people who are the subjects of this book were scientists, and what they did was science; only the saddest pedant would object to my use of the word science in this context.

Other things being equal, science should have taken off in Italy following the work of Galileo, and the revolution I describe in this book should have happened fifty years sooner than it did. But other things were not equal. The dead hand of the Catholic Church stopped the scientific revolution in Italy, with the conviction of Galileo for heresy serving as a powerful incentive for other scientists to give up their work, keep quiet about it, or head for more comfortable climes. In mainland Europe north of the Alps, the development of science was hampered by wars (often religious conflicts) and pestilence. Britain too had its civil war, but this was a different kind of conflict, and one which in the end produced a group of people determined, as we shall see, to keep religion out of science and to publish their discoveries for all to share. By comparison with most of its continental neighbours, by the second half of the seventeenth century England was a stable society (if partly because the horrors of the Civil War were still recent enough to serve as a reminder that a little tolerance might be better than the alternative), more or less democratic, increasingly prosperous, and tolerant, if not of all free-thinkers, at least of the kind of free-thinking involved in science. It was, indeed, fertile ground in which the seeds of science could grow. But although the plants that grew from those seeds began to flower in the 1660s, they were in fact planted in that fertile soil almost exactly at the beginning of the seventeenth century, and nurtured through the efforts of three men – William Gilbert, Francis Bacon, and William Harvey. They were the heralds of the scientific revolution in Britain, and it is with them that my story starts.

John Gribbin
March 2005

# BOOK ONE

# Heralds of the Revolution

*William Gilbert*

# I

# Scientific Minds

William Gilbert of Colchester deserves pride of place in any account of the scientific revolution of the seventeenth century, because he was the first person to set out clearly in print the essence of the scientific method – the testing of hypotheses by rigorous experiments – and to put that method into action. He did so to such effect that his discoveries in the field of magnetism were unsurpassed for two centuries; and by a happy calendrical coincidence, his great book on magnetism was published at the dawn of the new century, in 1600.

To put Gilbert's life and work into an historical perspective, in 1600 Elizabeth I was nearing the end of her long reign. The Spanish Armada had been defeated just twelve years earlier, and although the first attempt to plant an English colony in North America (at Roanoke Island, in what is now North Carolina) had failed in the mid-1580s, the successful attempt at establishing a permanent settlement in Jamestown, Virginia, would take place in 1606. In 1600, William Shakespeare was at the height of his creative powers, and in the 1599–1600 season the plays performed at his Globe Theatre in London were *Julius Caesar*, *Twelfth Night*, and *As You Like It*, while *Hamlet* and *The Merry Wives of Windsor* followed the next year. London itself was a city of some 75,000 inhabitants, with twice that number in the rapidly growing suburbs sprawling outside the city walls. It was filthy, smelly, and unhygienic; bubonic plague spread by the fleas that lived on rats often broke out in summer, when those that could afford to retreated to the countryside. The poor had no such option.

But the country as a whole was prosperous, and it was out of that prosperity, providing both the time and the means for a few gentlemen to indulge their interest in the world around them, that the scientific revolution grew. Gilbert's own background provides a classic example of the way growing numbers of wealthy people were becoming assimilated into society, nouveau riche families establishing their gentlemanly credentials and displaying at least some of the outward trappings of the old aristocracy.

The wealth of the Gilbert[1] family dated back only to the second half of the fifteenth century, when Joan Tricklove, the only daughter of John Tricklove, a wealthy merchant who lived in Clare, in the county of Suffolk, married one John Gilbert, who seems to have been an honest yeoman, probably uneducated, with no wealth of his own. Their son, William, became 'Sewer of the Bedchamber' to Henry VIII. We don't know what strings were pulled or palms greased to achieve this essentially meaningless position, but as a minor member of the Court this William Gilbert automatically became a gentleman. His own son, Jerome (one of several siblings), settled in Colchester, in Essex, and became Recorder of the town. He married twice. The first marriage, to Elizabeth Coggeshall, produced four sons (the eldest of whom was 'our' William Gilbert, born in 1544) and a daughter. After Elizabeth died, Jerome married Jane Wingfield, who presented him with five more children, including two sons, Ambrose and (confusingly) another William. Our William, as we shall see, rose to become one of the physicians to the Queen, whereupon a coat of arms was bestowed on his family, completing the rise of the Gilberts from yeoman stock to aristocracy, with vague hints of a mythical pedigree going back to William the Conqueror, in just three generations – a rise typical of hundreds of families under the Tudors.

Unfortunately, we know only the outlines of Gilbert's own life, because of two disasters that occurred later in the seventeenth century. Many of his papers and experimental equipment were left to the Royal College of Physicians when he died, and both their building and Gilbert's own former house in London were destroyed in the Great Fire of 1666. Any other material that might have been of interest to historians would have been at his house in Colchester, but that was

1. Spelling was something of a moveable feast in those days, and there are several variations in the form of the Gilbert family's surname (just as there are with Shakespeare's); we shall stick with 'Gilbert' for consistency, except where quoting original sources.

in a part of the town destroyed during the English Civil War, at what became known as the Second Siege of Colchester, in 1648. Indeed, it's a sign of how little we know about Gilbert's life that until the end of the nineteenth century the accepted date for his birth was 1540. This was based on the inscription on a monument erected to him by his two half-brothers, which says that he died in 1603 at the age of 63. Although it might have seemed reasonable to think that his brothers knew how old he was, a portrait of Gilbert painted during his lifetime (but now lost!) apparently had the date 1591 and gave his age as 48, implying a birth date in 1543. Since Gilbert personally gave the painting to the University of Oxford, it seems likely that the age given on it was more or less correct; and there is a manuscript in the Bodleian Library in Oxford which gives an actual birth date, 24 May 1544. We also know that Gilbert went up to Cambridge in 1558, and 14 would be about the right age for a scholar to enter the university in those days – 18 would be very unusual. So 1544 is now the accepted date of Gilbert's birth, and we assume a minor error on the portrait.[2]

Once Gilbert arrived in Cambridge, he began to leave traces which have survived to the present day. He stayed in residence at the university for eleven years, gaining both his BA (1560) and MA (1564), qualifying as a doctor (1569), and becoming a Senior Fellow (all at St John's College). There is no record of his activities between 1569 and 1573, and this has led to fanciful accounts that he may have travelled widely in Europe, perhaps even meeting Galileo. Since Galileo was only born in 1564, however, it seems unlikely that any such meeting occurred! The most likely explanation of Gilbert's activities during these years after leaving Cambridge and before setting up as a physician in London in 1573 is that it was just at this time that he became seriously interested in magnetism and carried out his early experiments; but the work certainly wasn't completed then. The best clue to when it was finished comes from a preface to Gilbert's book *De Magnete*, in which Edward Wright says that the work described in the book has been 'held back not for nine years only, according to Horace's Counsel, but for almost another nine'. It is an intriguing

2. See Roller. Full details of sources cited are given in Sources and Further Reading.

coincidence (if it is only a coincidence) that Gilbert first took an important office in the Royal College of Physicians, as Censor, in 1581; and it is hard to see how the successful and busy physician that Gilbert became after that time could have found time for much scientific work. If Gilbert began his key magnetic experiments (perhaps in Colchester) in the four years ending in 1573, then spent a decade using his spare time to refine his results, put the discoveries in order and prepared them for publication, that would exactly match Wright's comment. That places the key experiments, the first true application of what became the scientific method, in the early 1570s, in England, before Galileo was even 10 years old, and dates the completed work to the early 1580s, when Galileo was a medical student in Pisa.

The evidence that Gilbert started his medical practice in 1573 comes from references after his death (in 1603) that he had been a London physician for thirty years; but the first direct documentary evidence of his life in London dates from 1581, with that appointment as Censor, when he was already a member of the Royal College of Physicians. Throughout the rest of the century, Gilbert was a prominent member of the College, holding several offices including that of Treasurer from 1587 to 1594 and from 1597 to 1599, and being elected President in 1600. During this period he lived at Wingfield House (a property he may have inherited from his stepmother, whose maiden name was Wingfield), not far from the premises of the Royal College of Physicians, which were in Knightrider Street.

Part of the Gilbert legend has it that this house became the meeting place for a society of learned men, a kind of forerunner of the Royal Society. But the American historian Duane Roller has argued convincingly that this is a later misinterpretation based on references to the 'society' or 'college' at Wingfield House in sixteenth-century letters, which from the context of the times simply mean Gilbert's household. His career as a physician was crowned by the appointment as one of the Queen's doctors in 1601, and this too has been the subject of exaggerated interpretation down the years. Gilbert was just one of the royal panel of physicians, not singled out in any way, and he received the usual stipend of £100 a year (referred to as a 'pension') for his services. This has grown in the telling so that he is sometimes referred to as the Queen's Chief Physician (or her Personal Physician), and the pension becomes a mythical personal bequest in the Queen's will. None of this is true. But it is true that although Gilbert was confirmed in his post of Royal Physician when James I (in whose honour the first American colony was named) succeeded Elizabeth in 1603, he died later that year, almost certainly of the plague. He had never married, and family legend has it that he stayed a bachelor in order to provide for his younger half-siblings, which may explain why Ambrose and William the Younger erected a memorial to him. His lasting memorial, though, is the book *De Magnete*, published in 1600.

Although Gilbert's achievements, especially in developing the scientific method, were immense, it is astonishing to modern eyes just how much was already known about magnetism in the middle of the sixteenth century. Naturally occurring magnetic rocks, or lodestones, had been known since ancient times, both in China and in the Eastern Mediterranean. Our name 'magnet' comes from the old term *lithos magnetis*, or 'magnesian stone', which may have referred to lodestones found near the town of Magnesia, in Greece, although this is no more than a supposition. But although lodestones were known in ancient times, their properties were not investigated scientifically, and they were surrounded by superstition (as in the belief that they could cure illnesses) and exaggeration, as when Pliny tells us that:

> Near the river Indus there are two mountains, one of which attracts iron and one repels it. A man with iron nails in his shoes cannot raise his feet from the one or put them down on the other.[3]

There was also considerable confusion in ancient times about the relationship between what we now call electricity and magnetism. The Ancient Greeks valued amber, which they called *elektron*, and is actually the fossilized remains of resin from a variety of trees. They knew that when it was rubbed, it gained the power of attracting straws, small pieces of sticks, and even thinly beaten pieces of copper or iron. But they thought that the effect was a result of the amber

3. See *Pliny's Natural History*, trans. W. H. S. Jones, volume 6, Heinemann, London, 1949.

being heated by friction when it was rubbed,[4] and although they noticed that amber could attract all kinds of small objects, while lodestone could only attract iron, it still seemed to them that the attractive power of amber was essentially the same as the attractive power of the lodestone.

Nothing new was discovered about magnetism until the eleventh century AD, when suddenly we find references to the use of magnetic compasses in navigation. The origins of the discovery of this pointing property of magnets are lost, and it is clear that whoever first made the discovery realized its enormous commercial and military value, and kept it secret as long as possible. The discovery seems to have been made first in China, and a little later in Europe, assuming that the knowledge did not spread from China westward. One reason to think that the discoveries may have been independent of one another is that from the earliest references Chinese compasses are designed to point south, while European compasses are designed to point north. Near the end of the eleventh century, the Chinese Shen Kua tells us that:

> A geomancer rubs the point of a needle with the lodestone to make it point to the south, but it will always deviate a little to the east, and not show the south; that to use the needle, it may be put on water, but it would not be steady . . . that the best method is to hang it by a thread, and to prepare the contrivance, one has to single out a fine thread from a new skein of floss silk and fix it with a piece of beeswax on the middle of the needle, the latter to be hung up where there is no wind; that the needle would then always point to the south.[5]

Even in this early description, we see that people already knew that the compass needle did not point to the true south (or north) but deviated slightly. About a century later, the Englishman Alexander Neckam describes the magnetic compass in his books *De Utensilibus* and *De Naturis Rerum*, where he also explains how the magnetic needle can be balanced on a pivot. In the centuries that followed, a great deal of information, misinformation, and superstition grew up

4. This was not a completely mad idea. The Greeks had noticed that small pieces of straw and so on are 'attracted' to a fire (we now know this is due to convection currents set up in the air) and thought that something similar might be happening with the amber.

5. Quoted by Roller.

as a result of interest in the compass. As it became appreciated that the lodestone had two special points, just as the heavenly sphere has two special poles, even before it was widely understood that the Earth is round it was common to make lodestones round, to mimic the shape of the heavenly sphere. Such a sphere has two magnetic poles, one of which points north and the other south. It was known that with two such lodestones opposite poles attract and similar poles repel one another, and that if a lodestone is cut in half between the two poles, two new poles appear at the cut, each the opposite of the original pole in that half of the lodestone. It was even understood that you could find the poles of a spherical lodestone by laying bits of iron wire on the surface of the sphere and seeing how they align to point to the magnetic poles. But there was confusion about why a compass needle should point north – was it because it was attracted by the Pole Star, or was it because there was a magnetic island far to the north of Europe? Or was it just because it was in the nature of magnetic needles to point north? And although a magnetized needle would *point* to the north, it did not try to *move* to the north, even though it would both point to and move towards a lodestone placed near it.

Magnetism was also still seen as having medicinal properties, and a supposed cure for gout, for example, was to bandage a piece of magnetic material tightly up against the affected limb. This at least had the merit, unlike some medieval medical treatments, of doing no harm to the patient. Yet another key property of the compass needle, the way it dips to point slightly below the horizontal, was only discovered in the sixteenth century, just about in Gilbert's lifetime. Although the dip was mentioned in a letter written by the German Georg Hartmann in 1544, this was not published at the time, and the first report of the discovery to reach a wide audience came from the London-based instrument maker Robert Norman, in 1581. This discovery essentially completed the package of information about magnetism (and electricity) that Gilbert set out to explain and understand through his experiments, and to describe in his great book.

We don't intend to take you through all of Gilbert's work, because the important point we wish to emphasize is not *what* he discovered, but *how* he discovered it. The full title of his masterwork is *De Magnete Magneticisque Corporibus, et de Magno Magnete Tellure;*

GVILIELMI GIL
BERTI COLCESTREN-
SIS, MEDICI LONDI-
NENSIS,

DE MAGNETE, MAGNETI-
CISQVE CORPORIBVS, ET DE MAG-
no magnete tellure; Phyſiologia noua,
*plurimis & argumentis, & expe-*
rimentis demonſtrata.

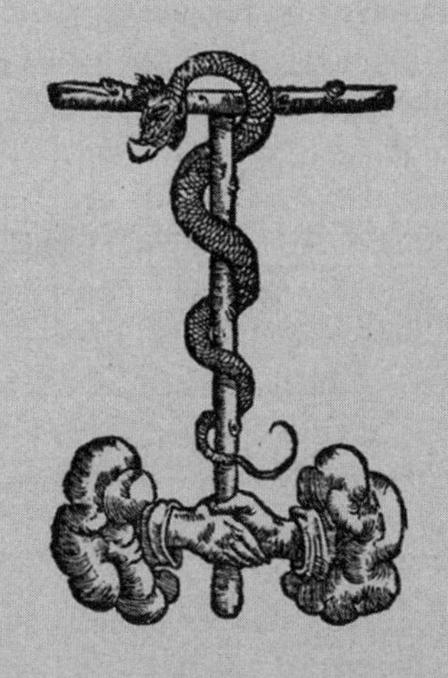

LONDINI

EXCVDEBAT PETRVS SHORT ANNO
MDC.



*Title page from* De Magnete

*Physiologia Nova, Plurimis et Argumentis et Experimentis Demonstrata*, which is usually translated as *On the Loadstone and Magnetic Bodies and on the Great Magnet the Earth*, with the subtitle ignored. Gilbert sets out his stall as the practitioner of a new kind of investigation of the world in the preface of his book, pulling no punches as he kicks off with the assertion that:

In the discovery of hidden things and in the investigation of hidden causes, stronger reasons are obtained from sure experiments and demonstrated arguments than from probable conjectures and the opinions of philosophical speculators of the common sort . . .

at once distancing himself from the school of natural philosophy, dating back to the Ancient Greeks, which attempted to unravel the mysteries of the Universe solely by thinking about them, without actually carrying out experiments. It's worth quoting extensively from that preface, to make it clear that not only was Gilbert doing something new, he was well aware of the revolutionary nature of his new style of investigation:

Every day, in our experiments, novel, unheard-of properties came to light . . .

But why should I, in so vast an ocean of books whereby the minds of the studious are bemuddled and vexed – of books of the more stupid sort whereby the common herd and fellows without a spark of talent are made intoxicated, crazy, puffed up; and are led to write numerous books and to profess themselves philosophers, physicians, mathematicians, and astrologers, the while ignoring and contemning men of learning – why, I say, should I add aught further to this confused world of writings, or why should I submit this noble and (as comprising many things before unheard of) this new and inadmissible philosophy to the judgment of men who have taken oath to follow the opinions of others, to the most senseless corrupters of the arts, to lettered clowns, grammatists, sophists, spouters, and the wrong-headed rabble, to be denounced, torn to tatters and heaped with contumely. To you alone, true philosophers, ingenuous minds, who not only in books but in things themselves look for knowledge, have I dedicated these foundations of magnetic science – a new style of philosophizing. But if any see fit not to agree with the opinions here expressed and not to accept certain of my paradoxes, still let them note the great multitude of experiments and discoveries – these it is chiefly that cause all philosophy to flourish; and we have dug them up and demonstrated them with much pains and sleepless nights and great money expense. Enjoy them you, and, if ye can, employ them for better purposes. I know how hard it is to impart the air of newness to what is old, trimness to what is gone out of fashion; to lighten what is dark; to make that grateful which excites disgust; to win belief for things doubtful; but far more difficult is it to win any standing for or to establish doctrines that are novel, unheard-of, and opposed to everybody's opinions. We care naught, for that, as we have held that philosophy is for the few.

We have set over against our discoveries larger and smaller asterisks according to their importance and their subtility. Let whosoever would make the same experiments handle the bodies carefully, skillfully, and deftly, not heedlessly and bunglingly; when an experiment fails, let him not in his ignorance condemn our discoveries, for there is naught in these books that has not been investigated and again and again done and repeated under our eyes. Many things in our reasonings and our hypotheses will perhaps seem hard to accept, being at variance with the general opinion; but I have no doubt that hereafter they will win authoritativeness from the demonstrations themselves . . .

This natural philosophy (*physiologia*) is almost a new thing, unheard of before; a very few writers have simply published some meagre accounts of certain magnetic forces. Therefore we do not quote the ancients and the Greeks as our supporters, for neither can paltry Greek argumentation demonstrate the truth more subtilly nor Greek terms more effectively, nor can both elucidate it better . . .

To those men of early times and, as it were, first parents of philosophy, to Aristotle, Theophrastus, Ptolemy, Hippocrates, Galen, be due honour rendered ever, for from them has knowledge descended to those that have come after them: but our age has discovered and brought to light very many things which they too, were they among the living, would cheerfully adopt. Wherefore we have had no hesitation in setting forth, in hypotheses that are provable, the things that we have through a long experience discovered.[6]

6. All quotations from Gilbert are from the Fleury Mottelay translation; see Sources and Further Reading.

In fact, although not appealing to the authority of the ancients, Gilbert then proceeds to give a clear account of the history of investigations of magnetism, before proceeding to his own discoveries. A couple of examples will suffice to highlight the way in which he worked, and why he can be regarded as the first scientist.

Gilbert's first objective is to draw a distinction between magnetism and the amber effect (for which he introduces the term electricity) in order to clear the air before moving on to his study of magnetism itself. In order to do this, he has to carry out a thorough investigation of electricity, and to help him he invents the first electroscope (he called it a versorium), in the form of a light needle, made of metal, 'three or four fingers long' and 'poised on a sharp point after the manner of a magnetic pointer' (that is, a compass needle). When a piece of rubbed amber, or other suitable material, is brought near to one end of the needle, the pointer revolves. Using this sensitive detector, Gilbert set out to investigate the properties of electricity. The Greeks had speculated that the attraction might be caused by the warmth of rubbed amber, and in all the time since them this had remained a possible explanation of the phenomenon. It was Gilbert who took what seems to us the obvious step of warming amber by other means, and finding that this does not produce an attraction (nor, as he pointed out, do other warm objects display electric attraction). It was the rubbing, not the warmth, that mattered.[7] Gilbert suggested that the rubbing removed a 'humour' from the body, and left behind an 'effluvium' surrounding the rubbed object; if you replace these terms by, respectively, 'charge' and 'electric field', this is essentially the modern view of what is going on. In all, there are 33 discoveries denoted by asterisks in the chapter of *De Magnete* on electricity, indicating that Gilbert had carried out all these experiments for himself. Although some of these discoveries might have pre-dated him, there is no surviving record of any earlier work on any of the 33 discoveries, which range from the discovery that a wide variety of other materials (such as sapphire, sulphur, and sealing-wax) attract light objects when

7. In the same spirit, Gilbert later tests the old wives' tale that garlic will demagnetize iron or a lodestone by actually rubbing them with garlic and showing that there is no effect. But this didn't stop the old wives' tale persisting through the seventeenth century!

rubbed to the fact that solar heat concentrated on to amber with a concave mirror does not result in attraction, to the fact that an electric object attracts small pieces of material towards itself in a straight line. And, of course, the electric force, as we would now call it, attracts a wide variety of materials, not just iron. In all this work, and in the theoretical explanations of the things he observed, Gilbert single-handedly established electricity as a new branch of science, distinct from magnetism. Virtually nothing was added to his work in the seventeenth century, so that it provided the jumping-off point for the eighteenth-century work which led to the concept of electric charge.

In his work on magnetism, Gilbert used the spherical lodestones that we have already described, which he called terrellae, meaning 'little Earths'. It was at the very heart of his magnetic philosophy that he regarded these as models of the Earth itself, and that he thought of the Earth as a giant spherical magnet. This is an important distinction. Earlier investigators made their lodestones spherical in order to mimic the shape of the heavens; Gilbert made his spherical in order to mimic the shape of the Earth. The nature of the magnetic influence of a terrella was investigated using a magnetized compass needle, which would align itself to point to the poles of the terrella just as the terrella would align itself to point to the poles of the Earth, with no need to invoke magnetic islands to the north of Europe, or some influence from the Pole Star. In these experiments, Gilbert was the first person to appreciate that, because magnetic opposites attract, it is the *south* pole of a magnet that points to the *north* pole of the Earth; in modern language, we sometimes refer to the 'north-seeking' pole of a compass needle to make the point clear. As Gilbert puts it:

All who hitherto have written about the poles of the loadstone, all instrument-makers, and navigators, are egregiously mistaken in taking for the north pole of the loadstone the part of the stone that inclines to the north, and for the south pole the part that looks to the south: this we will hereafter prove to be an error. So ill-cultivated is the whole philosophy of the magnet still, even as regards its elementary principles.

This is not just some matter of hair-splitting semantics, or a cheap gibe at his predecessors. The point is that Gilbert recognizes that it is the same process that makes a magnet that is free to move orient itself relative to a fixed magnet that makes a compass needle orient itself with respect to the Earth's magnetism. The Earth *is* a magnet, and therefore understanding magnetism will help us to understand the Earth. This is the first example of trying to understand global (ultimately, universal) forces by carrying out experiments on a laboratory scale. And, as Gilbert appreciates, magnetism can then be used to provide information about what is going on deep inside the Earth, in regions we can never see.

The other aspect of Gilbert's work that we particularly want to draw attention to takes us in the other direction – not inwards to probe the structure of the Earth, but out into space. Gilbert was among the first to appreciate that there is something more to magnetism than an attractive influence, or force; and he was the first to set out clearly just what that 'something' seemed to be. In doing so, he came very close to the modern idea of a field, suggesting that the magnetic effect surrounded the Earth (and, by implication, other planets) in a sphere of influence. The genesis of this idea also reminds us that Gilbert was not an isolated genius working alone, but a man of his times, in touch with scientific investigations being carried out by others. In the last decades of the sixteenth century, the time was ripe for the kind of investigation of magnetism that he carried out, and if he had not done it then somebody else surely would have before too long.

This work jumped off from the investigation of magnetic dip, which was first fully described in print by Robert Norman, in a little book called *The New Attractive*, which he published in 1581, several years after what seems to have been Gilbert's most productive experimental period. Since Gilbert repeated some of Norman's experiments and described their results in *De Magnete*, he must have found some time during his busy life in the 1580s to do at least a little scientific work; it is also possible that some of his experiments on dip pre-dated Norman's work, although they were not published until 1600. We shall never know, and it doesn't really matter.

Norman tells us that he began to investigate magnetic dip (he called it 'declination', but that term has a quite different meaning today) when he got angry at the problems it caused him when he was manufacturing compasses. It was well known by then that a magnetized needle suspended from its mid-point would not lie horizontally, but with the north-seeking end pointing downwards, below the horizon.[8] In order to compensate for this, instrument makers such as Norman had to snip a little piece off the north-seeking end of the needle so that it would balance perfectly on its pivot and make the compass usable for navigation. Norman became so cross when he spoiled a particularly fine compass by cutting too much off the needle that he decided to find out just why the magnetized needles behaved in this way, and to do so he invented a new kind of compass, the dip circle. In a dip circle, a graduated circular rim (like the tyre of a bicycle wheel) is set up vertically, and a compass needle is supported on an axle in the middle of the wheel so that it can rotate freely in the vertical plane. Norman found that needles set up in this way always pointed downward at the same angle, 67°, in London, and he conjectured that the angle of dip might be related to latitude. But his great insight, the idea that Gilbert picked up on and developed, was that the magnetic needle is not being *attracted* towards the North Pole; it simply *points* to the North Pole, indicating the direction of something (which we would now call the magnetic field) in the vicinity of, in this case, London. He said that 'In my judgment [the point attractive] ought rather to be called the point respective'. Norman reinforced this conclusion with a particularly subtle experiment, which Gilbert repeated and described in *De Magnete*. He took a piece of iron or steel a couple of inches long, and thrust it through a piece of cork. He then filled a glass vessel with water, and by painstakingly carving away at the cork little by little, made the needle float horizontally in the water, a few inches below the surface, suspended by the buoyancy of the cork, 'like to the beam of a pair of balances being equally poised at both ends'. Then:

> Take out the same wire without moving the cork, and touch with the [lodestone], the one end with the south of the Stone, and the other end with the north, and then set it again in the water, and you shall see it presently turn upon his own centre, showing the aforesaid declining property, without descending to the bottom, as by reason it should, if there were any attraction downwards.[9]

8. In the Northern Hemisphere. In the south, it is the south-seeking end that dips below the horizon.

9. Quoted by Harré. Other quotes from Norman are from the same source.

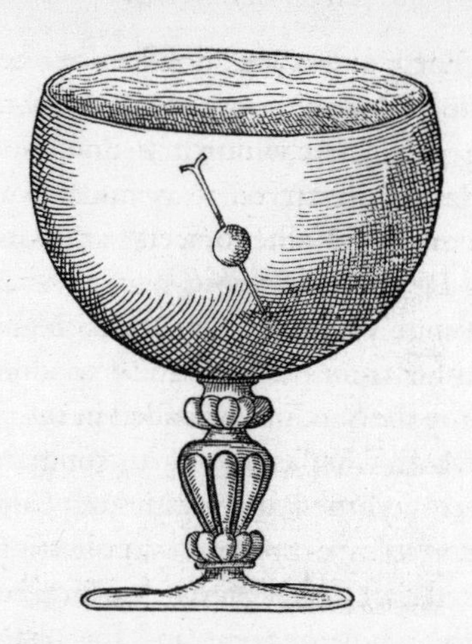

*The experiment performed by Robert Norman and repeated by Gilbert to show that magnetism is directional*

So the needle is lining up with what we would now call the magnetic field, but which Norman refers to as a 'virtue'. He says, 'I am of opinion, that if this virtue could by any means be made visible to the eye of man, it would be found in a spherical form extending round the Stone.'

Gilbert was able to go further, by investigating the way dip varied around his terrellae – remember that it is a key contribution to scientific thinking that he regarded these models as miniature Earths, and that he could therefore extrapolate from their behaviour to the behaviour of the real Earth. He was able to show that the angle of dip does indeed vary with latitude, and he found another way of showing that what is being measured is a direction, not an attraction, by demonstrating that for a spherical terrella the dip was always the same at any particular latitude, whatever the strength of the lodestone. If the dip were due to an attraction, you would expect it to be more pronounced if the magnetism of the stone were stronger, but this is not the case. The needle takes up its orientation relative to the terrella (or the Earth) as a *whole*, not because of the strength of an attraction towards the pole. 'This movement,' says Gilbert, 'is produced not by any motion away from the horizon towards the earth's centre, but by the turning of the whole magnetic body to the whole of the earth.'

This idea of the Earth extending an influence out into space around itself links with Gilbert's speculations about the nature of the Universe itself and the place of the Earth within it. In doing so, he makes another conceptual leap. Having used terrellae as models for the Earth, he now uses the Earth as a model for other objects in the Universe. Copernicus had published his *De Revolutionibus* only in 1543, the year before Gilbert was born; and as the fate of Galileo highlights, in Gilbert's lifetime it was still far from being received wisdom that the Earth is just a planet orbiting the Sun, unsuspended in the void. Indeed, it was still a matter of debate whether the Earth rotated on its axis, or the heavens revolved around the Earth – although Gilbert left his readers in no doubt about where he stood on that question.

As Gilbert pointed out, 'either the earth whirls in daily motion from west to east, or the whole heavens and the rest of the universe of things necessarily speeds about from east to west.' But the stars are so distant from us that they would have to travel at enormous speeds to complete the circuit in twenty-four hours. He dismissed the idea of the 'adamantine spheres' out of hand – 'nor is there any doubt that even as the planets are at various distances from earth, so, too, are those mighty and multitudinous luminaries [the stars] ranged at various heights and at distances most remote from earth, [situated] in thinnest aether, or in that most subtile fifth essence, or in vacuity.' In which case, how could they all keep in step as they wheeled round the Earth?

> Astronomers have observed 1022 stars; besides these, innumerable other stars appear minute to our senses; as regards still others, our sight grows dim, and they are hardly discernable save by the keenest eye; nor is there any man possessing the best power of vision that will not, while the moon is below the horizon and the atmosphere is clear, feel that there are many more indeterminable and vacillating by reason of their faint light, obscured because of the distance. Hence, that these are many and that they never can be taken in by the eye, we may well believe. What, then, is the inconceivably great space between us and these remotest fixed stars? and what is the vast immeasurable amplitude and height of the imaginary sphere in which they are supposed to be set? How far away from earth are those remotest of the stars; they are beyond the reach of eye, or man's devices, or man's thought. What an absurdity is this motion [of spheres].

Gilbert seems to be on the edge here of describing an infinite universe filled with stars; indeed, he states that 'there cannot be diurnal motion of infinity or of an infinite body'. Tellingly, he then adds another kind of argument to his list of reasons for believing that the Earth rotates, asking those who still believe that the heavens rotate around the stationary Earth, 'what structure . . . can be imagined so strong, so tough, that it would not be wrecked and shattered to pieces by such mad and immeasurable velocity?' This is a question that simply would not have occurred to his predecessors. They regarded the heavens as something mystical, not subject to the same rules as solid objects here on Earth. But here is Gilbert, almost a century before Newton, implicitly assuming that the same laws of physics apply to the most distant stars as to a lump of matter on Earth. Science is encroaching on what used to be the territory of religion, with dire consequences for some scientists in some parts of Europe. It was still possible to be burned at the stake for expressing such views in Catholic countries; in England, you might be burned for being a Catholic, but not for offering a scientific opinion about the nature of the Universe. So Gilbert is free to say that 'the space above the earth's exhalations is a vacuum', and that 'the entire terrestrial globe, with all its appurtenances, revolves placidly and meets no resistance' in that vacuum. This is an implicit recognition of something studied in more detail by both Galileo and Newton – the idea of inertia, that an object once set in motion will continue to move as long as it is not affected by external forces (such as friction). Gilbert does not explicitly say that the Earth also moves around the Sun, but he refers approvingly to the work of Copernicus ('a man most worthy of the praise of scholarship') in the context of the motion of the other planets round the Sun. He tries, though, to explain all these motions in terms of magnetism, which was far from being unreasonable some eight decades before Newton published his theory of gravity, but which we will not dwell on here as it was just plain wrong.

*De Magnete* summarizes a body of work which marks the beginning of the application of the scientific method of investigating the world, and points the way for Galileo, Newton, and the other seventeenth-century pioneers. But it was fortunate for science, and for Gilbert himself, that the work was carried out and published in London rather than, say, Rome – just how fortunate we can see by looking briefly at the unfortunate fate of Gilbert's contemporary Giordano Bruno, and in more detail at the career of Galileo himself.

Bruno, who was just four years younger than Gilbert, was burned at the stake for heresy in 1600, the year *De Magnete* was published. He undoubtedly supported the Copernican idea of a Sun-centred Universe, and many accounts of his fate give the impression (or state specifically) that he died for his scientific beliefs. This is not quite the case. A key reason why Bruno was such an enthusiastic supporter of Copernicus was that he was a member of a sect that believed Christianity to be a corruption of the old Egyptian religion in which the Sun featured as a god, and he also rejected the idea of the Holy Trinity, arguing that Christ was created by God but was not God himself incarnate (a belief known as Arianism, after the teaching of the fourth-century Libyan Arius, who was excommunicated for his views in 321). Although the records of his trial have been lost, the evidence we have suggests that Bruno was specifically condemned for Arianism, not for being a Copernican. Nevertheless, the fact that Bruno *was* a Copernican and *had* been burned for heresy was distinctly discouraging for other Copernicans in Catholic Europe.

All of this, though, highlights a key point about the conflict between science and (some) religion at this time. The kind of enquiring mind that wondered about the nature of the Universe and the place of the Earth and humankind in the Universe was also the kind of enquiring mind that would wonder about the meaning of religion and of biblical texts. As well as being a Copernican, Bruno himself enthusiastically espoused the idea of an infinite Universe filled with an infinite number of stars. This idea seems to have originated with Thomas Digges, a sixteenth-century English astronomer (he died in 1595) whose father, Leonard, invented the astronomical telescope. The younger Digges almost certainly got his ideas about the Universe by looking at the Milky Way through a telescope, although the invention was kept secret at the time and was re-invented early in the seventeenth century. Digges published his cosmological speculations in 1576, and they probably influenced Gilbert's thinking about the Universe, as well as Bruno's. But Bruno, never one to do things by halves, seized on the idea that the Universe was filled with an infinite array of stars like the Sun, arguing that there must be other planets like the Earth orbiting those other suns, and other life-forms living on those planets. None of this endeared him to the Catholic Church; but all of it was the product of an enquiring mind, the kind of mind that can make a good scientist.

So Galileo Galilei, a 36-year-old professor of mathematics living in Padua, received two clear messages in 1600. From England came William Gilbert's great book on magnetism, setting out more clearly than ever before the way in which the scientific method could be used to obtain an understanding of the Universe. Galileo was so struck by this that he is said to have commented: 'I greatly praise, admire and envy this author, that a conception so stupendous should have come to his mind.' From Rome, uncomfortably closer to hand, came news of the fate of Giordano Bruno, spelling out the fact that the Catholic Church was not yet ready to allow enquiring minds free rein. Galileo's subsequent career involved a delicate balancing of the two, but when he did inevitably tilt the balance in the direction of science, his fate was at least less extreme than that suffered by Bruno.

Galileo Galilei is widely regarded as the founding father of the scientific investigation of the world. Although he added nothing to the understanding of the scientific method pioneered by Gilbert, it is appropriate that he should be held in such high esteem because he applied that method more widely, and was a more effective propagandist for science. Gilbert did it first, but his career as a physician and his focus on just one aspect of scientific investigation, magnetism, left the way clear for Galileo to be the first person to devote his life to what we would now recognize as science. In addition, ironically Galileo's run-in with the Church authorities in Rome, while stifling the progress of science in Italy, ensured widespread attention for his ideas and the new scientific method itself elsewhere.

Galileo owed both his first and his last name to an illustrious ancestor. The family name had been Bonaiuti until the middle of the fifteenth century, when it was changed to Galilei in honour of Galileo Bonaiuti, an eminent physician and magistrate in whose glory his descendants were eager to bask. 'Our' Galileo was born just over a hundred years after his eponymous ancestor died. The great Italian artist Michelangelo died in the same month that Galileo was born; later that year, 1564, William Shakespeare was born at Stratford-

*Galileo Galilei*

upon-Avon. Columbus had discovered America just seventy-two years previously, and Galileo would be 56 years old when the *Mayflower* carried the Pilgrim Fathers to New England.

The world was changing in the second half of the sixteenth century, but one aspect of the world had scarcely changed for more than a thousand years. In natural philosophy (the forerunner of science), the ultimate authority (indeed, virtually the *only* authority) was Aristotle, the Greek philosopher who had lived in the fourth century BC. Aristotle (who, among other things, was the tutor of the boy prince who grew up to become Alexander the Great) was an extremely influential figure in Greek philosophy, and later in Roman philosophy; but his pervasive influence in sixteenth-century Europe owed much to a chain of accidents.

After the fall of Rome, most of Aristotle's works were lost to European civilization, but survived in Byzantium, capital of the Eastern Roman Empire, from where the Greek texts were translated first into Syriac, then from Syriac into Arabic, brought back to the attention of European scholars by the spread of Islam (all the way into Spain by the eighth century AD), and translated from Arabic into Latin in the twelfth and thirteenth centuries. As you can imagine, the Latin versions were very scrambled after all these wanderings, but the interest they aroused led to the rediscovery of the original Greek texts in Byzantium (by then known as Constantinople, and today as Istanbul) and to their translation from the Greek direct into Latin, the universal language of European scholars in those days, at the instigation of Thomas Aquinas.

All these wanderings were, in a way, appropriate. Aristotle himself used to teach his students while walking about, and as a result his followers (who had to be literally followers in those days) became known as 'Peripatetics', a name which was still used in Galileo's time. Unfortunately, far from being the free and easy kind of discipline that such a name conjures up, Aristotelianism was both rigid and wrong.

The really bad thing about Aristotelianism was that it was based on the notion that the truth about the world could be determined by pure thought – philosophy – without actually carrying out tests (what we would now call experiments) to see if the theories and hypotheses were right. Thus, for example, according to the Peripatetics (but translating into modern units), not only did a heavier object fall faster than a lighter object, but, specifically, an object weighing 100 kilos would fall 100 metres in the same time that it takes for an object weighing 1 kilo to fall 1 metre, both of them starting from rest. The blindingly obvious fact that this could be tested by dropping two objects with different weights at the same time simply did not come into it.

Aristotle believed that the Sun and stars (and other heavenly bodies) were perfect and unchanging (among other things, this meant they were perfectly spherical), that the Earth was at the centre of the Universe, and that the Sun, Moon, planets, and stars were carried round the Earth in perfect circles by crystalline spheres, to which they were attached. None of this was really crazy in the fourth century BC, but it became set in stone because the Peripatetics simply refused to test their ideas through doing experiments, or to take any notice of the results of other people's experiments. Worse, when those Greek texts were translated into Latin by the monks working under the orders of Thomas Aquinas, they struck a chord with the Catholic Church. Aristotle's idea of *geometrical* perfection (which had nothing at all to do with religion) harmonized, in the eyes of Thomas Aquinas, with the idea of the perfection of God and of God's works. So the whole kit and caboodle of Aristotelian cosmology (modified only by the improvements made by Claudius Ptolemy in the second century AD, who introduced the idea of 'epicycles', with each planet orbiting in a tiny circle, the centre of which orbited the Earth) was swallowed up wholesale by the Catholic Church, with the implication that questioning any of this stuff was questioning the perfection of God, and therefore heresy.

This was the situation when Galileo came on the scene. Although cheaply printed books (and pamphlets) had begun to be available early in the sixteenth century, spreading knowledge outside the traditional centres of the monasteries and the Church-dominated universities, and thinkers such as William Gilbert, Nicolaus Copernicus, and Tycho Brahe had begun to question the received wisdom (from their relatively safe positions in northern Europe, not too close to Rome), in the formal centres of learning the party line stood firm. The role of the professors of natural philosophy (as science was then called) was not seen as one of questioning the Universe to find out how it worked, but of preserving the Aristotelian tradition, and making sure that students learned it in just the same way that the professors had learned it when they were students.

But Galileo had a head start on the other students. His father, Vincenzio, was an accomplished professional musician (a respected calling then) and keenly interested in mathematics and musical theory. In his *Dialogue of Ancient and Modern Music*, published just when Galileo went off to university, he said:

> It appears to me that they who in proof of any assertion rely simply on the weight of authority, without adducing any argument [that is, experimental evidence] in support of it, act very absurdly.

Nothing could be more anti-establishment and anti-Aristotelian. Vincenzio could get away with it in the world of music, though even there his ideas caused a stir. His son, brought up to question authority and find things out for himself, would find it harder to get away with such ideas when it came to discussing the place of the Earth in the Universe.

Galileo was born on 15 February 1564, in Pisa, in the Tuscany region of northwest Italy, a centre of the Renaissance. The town, ruled by the Duke of Florence, Cosimo de' Medici, was flourishing and prosperous, and Cosimo himself was to be crowned Duke of Tuscany by the Pope in 1570, as a reward for his military campaigns against the Moors.

Galileo was the eldest of seven children (his father, Vincenzio, and mother, Giulia, had married in 1562), but the only ones we know anything about are his sister Virginia (born in 1573), brother Michelangelo (born in 1575), and another sister, Livia (born in 1587). The others, a boy and two girls, may have died in infancy. Vincenzio came originally from Florence (where he had been born in 1520), from a respected family now slightly less affluent than in previous generations, but still distinctly part of respectable society. In 1572, Vincenzio went back to Florence to re-establish himself there; his wife and younger children went with him, but Galileo stayed in Pisa with one of his mother's relatives for a while, and joined them in 1574, when he was 10. It may be that there was some plan for Galileo to learn the wool trade, which was the relative's business; if so, nothing came of it.

In Florence, the family moved in high circles; although never wealthy, Vincenzio was a Court musician and mixed with dukes and princes.

And Florence was the capital of Tuscany, the very centre of the Renaissance, then at the height of its fame as the intellectual heart of Europe.

Until he was 11, Galileo was educated privately at home, by his father and the occasional tutor. Among his other talents, Galileo developed an aptitude for the lute (his father's favourite instrument) and reached professional standard; although he only played for amusement, the instrument remained a source of pleasure throughout his life. In 1575, the boy was sent away for more formal education, at the monastery at Vallombroso, in the mountains 30 kilometres east of Florence. He liked the way of life there so much that at the age of 15 he joined the order as a novice. His horrified anti-establishment father extracted him from the clutches of the monks on the pretext of taking him to see a doctor about an eye infection. The infection was genuine, and so was the doctor, but what Vincenzio didn't mention was that he had no intention of bringing the boy back. Although Galileo's studies continued in Florence under the supervision of the same monkish order, he lived at home where his father could keep him clear of any indoctrination.

Vincenzio was keen to see his eldest son established in a respectable and financially rewarding career, and intended him to become a physician. So in 1581 (when Gilbert had long since completed most of his magnetic studies, and was well established as a respected physician in London) Galileo was enrolled as a medical student at the University of Pisa, where he could reduce his expenses by living with the same relative, Muzio Tedaldi, who had looked after him for a couple of years when he was a child. Almost immediately, he established a reputation as an awkward student who was not afraid to question established ideas, and earned the nickname 'the wrangler'. Much later in life, he wrote of his first formal encounter with Aristotelian ideas, and how he had immediately thought of a refutation of the idea that objects with different weights fall at different speeds. He had seen with his own eyes that hailstones with very different sizes and weights reach the ground together in a storm, and according to the Peripatetics that could only mean that the heavier hailstones had started their journeys higher in the clouds – not just higher, but *exactly* the right amount higher to ensure that they arrived at the ground with their lighter counterparts. To the young Galileo it made more sense to argue that all the hailstones had started in the same cloud layer, and fallen at the same speed.

But this wasn't what he was supposed to be concerned with, and had nothing to do with the study of medicine. His career was turned around early in 1583, during his second year of medical studies, when Galileo met Ostilio Ricci, the Court Mathematician of the Grand Duke of Tuscany. The whole Court took up residence in Pisa each year from Christmas to Easter, which was how Galileo got to know Ricci socially. It was only by accident that he called on his new friend when he was giving a lecture to some students, and stayed to listen, fascinated by the subject. It was Galileo's first contact with mathematics (as distinct from basic arithmetic), and he was hooked. He sat in on other lectures by Ricci, and began to study Euclid instead of medicine.

Ricci was sufficiently impressed by Galileo's quick grasp of the subject to back him up when he asked Vincenzio for permission to give up his medical studies and become a mathematician. Knowing how limited the job opportunities for mathematicians were, Vincenzio refused; but Galileo continued to study mathematics anyway. The result was that he left the university in 1585 without a degree of any kind, and went back to Florence, where he tried to make a living by giving private tuition in mathematics and natural philosophy.

It was while he was a student in Pisa that Galileo probably made one of his most profound observations. Legend has it that it was while watching a swinging chandelier in the cathedral, during a rather dull sermon, that Galileo realized that the pendulum always took the same time for one swing, whether it swung through a long arc or a short one (he timed the swing using his own pulse). The popular story further says that he immediately rushed home, carried out a series of experiments with pendulums of different lengths and weights, and invented a device called a pulsilogia, which doctors could use for timing the pulse of a patient.

Most of the story is not true; the detailed experiments were carried out later (in 1602), and the pulsilogia was then invented by a friend of Galileo, using his discovery. But late in life Galileo told the story of watching the swinging chandelier to Vincenzo Viviani, who became Galileo's scribe after the great man went blind, and Viviani wrote the first biography of Galileo, which always shows him in a good light, but is not always entirely accurate.

Although Galileo struggled to make a living, he established a reputation as a natural philosopher and began to carry out experiments and write down his ideas about the nature of the world, though he did not publish them at this time. In order to make progress in an academic career in those days, it was essential to have an influential patron, and Galileo found one in the form of the Marquis Guidobaldo del Monte, an aristocrat with a keen interest in science (and not just a dilettante; he had written an important book on mechanics). Partly thanks to del Monte's influence, in 1589 Galileo became Professor of Mathematics at the University of Pisa – the same university he had left, without a degree, just four years earlier.

Mathematics was not regarded as a very important subject in those days, and the salary associated with this job was low compared with that of other professors – a mere 60 crowns a year, compared with the 2,000 crowns received by the Professor of Medicine. But it was a start. Galileo augmented his income by private tuition – not just giving the benefit of his knowledge to students for the odd hour at a time, but taking in lodgers, the sons of the rich and powerful, who lived in his house and received the benefit of contact with the teacher at all hours. This was a standard procedure, but one which not only helped Galileo to make ends meet, but spread his fame as a teacher and thinker widely as the young men eventually finished their courses and went back to their homes, some of them in distant lands.

But although Galileo was a popular teacher, he did not get on with the establishment in Pisa. He led a curious double life, diligently teaching the Aristotelian texts that were the backbone of his official courses at Pisa, but questioning the Aristotelian way of thinking more and more in private, and not always keeping these thoughts to himself. He discussed these ideas with his students, argued with other professors, and wrote the first draft of a book on motion, *De Motu*, although at the last minute he decided not to publish it (probably a wise decision, since this early work is a mixture of Aristotelian and Galileo's own ideas; unlike the polished book published under the same title several years later).

The most famous story about Galileo's time in Pisa is another of those legends which have only a tenuous connection with reality – the story that he dropped cannonballs of different weights from the Leaning Tower to show his students that they reached the ground together. Once again, the source of the story is Vincenzo Viviani, who was only 17 when he began working for Galileo, and clearly hero-worshipped the old man. Like other stories in Viviani's biography, it is based on Galileo's reminiscences, when Galileo himself was in his seventies.

Although the idea was certainly in the air at the time – in 1586, a Flemish engineer, Simon Stevin, had carried out experiments dropping lead weights from a height of about 10 metres, and had published his results – Viviani has got this confused with an occasion in 1612, when one of the professors associated with the traditional Aristotelian world view tried to refute Galileo's arguments by dropping different weights from the tower, intending to show that they reached the ground at different times. They didn't *quite* reach the ground simultaneously, which the Aristotelians claimed as a refutation of Galileo's argument; but Galileo's response summed up the salient features of the outcome of the experiment:

> Aristotle says that a hundred-pound ball falling from a height of a hundred cubits hits the ground before a one-pound ball has fallen one cubit. I say they arrive at the same time. You find, on making the test, that the larger ball beats the smaller one by two inches. Now, behind those two inches you want to hide Aristotle's ninety-nine cubits and, speaking only of my tiny error, remain silent about his enormous mistake.

Even three decades before those events, along with his scientific free-thinking, Galileo, still only in his twenties in the late 1580s, openly scoffed at the pomposity of the professors in Pisa, ridiculing their official uniform of the toga. He enjoyed life to the full, including, literally, wine, women, and song, plus good food, until the end of his days; many of his surviving letters to or from friends in distant places mention the gift of a barrel of wine. Well before his three-year appointment came to an end, it was clear that Galileo didn't really fit in at the University of Pisa, and that the appointment would not be renewed. In any case, Galileo had his eyes set on higher things, and began campaigning, with the aid of his patrons, to capture the more prestigious and better-paid Chair of Mathematics at the University of Padua.

The attempt was made more urgent when his father died in 1591, making Galileo the head of the family, and burdening him with heavy debts. Before he died, Vincenzio had promised a substantial dowry to his daughter Virginia, and Galileo and his brother Michelangelo became responsible for this; unfortunately, far from paying his share, Michelangelo, who became a musician, would frequently come to Galileo for financial support for the rest of his own life.

Padua was part of the Venetian republic, a rich and powerful state that had one of the most enlightened governments in Italy, tolerating all kinds of ideas that were frowned upon by Rome. In order to press his claims to the Chair of Mathematics in Padua (which is about 20 miles from Venice itself), Galileo travelled to Venice, where he was helped by the Tuscan Ambassador. He put on an impressive display of his ability and charm, making influential new friends, including Gianvincenzio Pinelli, finding a new patron in the form of General Francesco del Monte (the younger brother of Guidobaldo), and securing the appointment for four years (at a salary of 180 crowns a year), with an option for the head of the republic, the Doge, to renew it for two years if he wished. All that Galileo needed now was the permission of the then Grand Duke of Tuscany, Ferdinando, to take up this appointment with a foreign state. It seems to have been granted without a second thought, and Galileo took up his new post in October 1592. He was 28 years old, and was to spend eighteen years as Professor of Mathematics in Padua; he later said that these were the happiest years of his life.

The liberal-minded intellectual community of Padua extended beyond the university, and one focus of its activity was the house of Galileo's new friend, Pinelli, where he stayed until he could find a suitable house of his own. There, he was able to use Pinelli's large library, and he met people who would figure strongly in his life in later years, including Paolo Sarpi and Robert Bellarmine.

In public life, Galileo made his name with a treatise on military fortifications, a subject of keen interest and considerable importance to the Venetian state, which maintained its position of strength as much through military power as through trade. This was quickly followed by a book on mechanics, based on the lectures he was giving in Padua. Bizarrely to modern eyes, in those days it was still widely believed – and taught – that a large weight could be lifted by a small force, with no tradeoff. As Galileo's book developed through several editions, he spelled out clearly that this was not possible, and that you can't get something for nothing. In systems involving levers and pulleys what is really happening, he explained, is that a small force moving through a large distance is shifting a large weight through a small distance. In a neat analogy, he explained that this was as if the large weight was divided up into a set of smaller weights, which were shifted in turn by the small force – so, for example, in a pulley system where a 1 kilo weight lifts a 10 kilo weight by 1 metre, the 1 kilo weight moves downward through 10 metres, just as if it had made ten journeys each 1 metre long to lift ten single 1 kilo weights through 1 metre each.

Money was constantly a problem to Galileo during his time in Padua, and one of the recurring themes of his work was to come up with an invention that would make him rich – or at least, rich enough to meet his obligations to his family. One of his inventions, around 1593, was an early kind of thermometer, a glass tube with a bulbous swelling in one end, and open at the other. The tube was heated to expel air, and then placed upright in a bowl of water, with the open end under the water. Water rose up the tube as the remaining air in it cooled and contracted. Then, like an upside-down version of the modern thermometer, when it was warmer the air in the bulb would expand, pushing the water level downward. It was ingenious, but not very practical (not least because the water level in the tube depended on air pressure, as well as temperature) and made no money. Another invention – successful technically but never a commercial success – was a system for lifting water for irrigation.

In 1595 or 1596, though, Galileo came up with a modest winner, a device known as a 'compass', but actually a kind of all-purpose calculating instrument, a forerunner of the slide rule that was an essential scientific implement in the days before pocket calculators. Galileo's compass started life as a device for gunners to use in calculating elevations when ranging their guns. It was a distinct improvement over the existing system, which involved the gunners standing out in front of their guns and hanging a device rather like a carpenter's set square in its mouth – not the safest place to be in the heat of battle (incidentally, the elevations on this older form of gauge were marked off in intervals of 7.5 degrees of arc, called 'points'; the lowest point, for zero elevation, was blank, so the term 'point blank' came to refer to very short-range gunnery).

Galileo's gunnery aid was developed into an all-purpose surveying instrument, and by 1597 Galileo had adapted it to include scales showing the relative densities of various metals, lines for calculating cube roots and square roots, and other data useful in geometrical applications. The instrument was such a success that for a brief period he had to employ a skilled artisan to make them; but he sold them relatively cheaply, and made his money out of the idea from the tuition fees he charged for teaching purchasers how to use them.

It was just as well that Galileo found his temporary new source of income at that time, because at the end of the 1590s and into the beginning of the seventeenth century his financial commitments increased. First, he formed an alliance with a local woman, Marina Gamba. They never married – they never even lived under the same roof – but formed a stable relationship which lasted more than ten years, and which produced three children, daughters in 1600 and 1601 and a son in 1606. Marina was a commoner, and such liaisons were themselves common, but kept fairly discreet in Italy in those days. Although the couple were clearly more than fond of each other, the social standing of their relationship was made quite clear when Galileo made his servants godparents to the children – although he loved the children dearly, and his son, Vincenzio, was later legitimized and became his heir.

In 1601, Galileo's younger sister, Livia, married, and together with Michelangelo he promised a substantial dowry. As before, Michelangelo never paid his share, and instead tried his luck as a musician first in Poland and then in Germany, each time with the aid of money 'borrowed' from Galileo and never repaid. Galileo kept himself afloat financially by taking in students, as he always had, getting advances against his salary, and borrowing money in his turn from rich friends, notably Giovanfrancesco Sagredo, a wealthy nobleman nine years younger than Galileo, a bachelor who could easily afford to help out his friend.

But don't think the financial difficulties stopped Galileo enjoying a full and happy life. He visited Venice regularly, and often went on trips into the hills near Padua, staying in the villas of his friends. One of these trips (probably in 1603) had unfortunate long-term consequences. On a hot day, after some healthy walking and a hearty meal, Galileo and his companions went to sleep in a room equipped with a kind of air conditioning, connected by ducts to cool caves. The ventilators leading to the caves were closed when the party went to sleep, but were then opened by a servant. Galileo and his two companions became severely ill as a result – so ill that one of them died, which suggests that something more than just the effects of cold, damp air on hot, overfed bodies was involved, and that some form of toxic gases may have come into the room from the caves along with the cool air. For the rest of his life, Galileo suffered repeated bouts of illness, which he blamed on this incident; the attacks left him confined to bed for weeks at a time with arthritic pains.

But none of this seriously affected his scientific work. It was in Padua that he carried out detailed experiments with pendulums, and studied acceleration by timing the motion of balls rolling down inclined planes – although very little of this would be made public until the 1630s. Always, Galileo carried out experiments himself to test hypotheses and find out how nature worked, rejecting the abstract 'philosophical' approach of the Peripatetics. Following the work of Gilbert, which he knew about from *De Magnete* and discussed approvingly in correspondence with his friends, including Sagredo and Sarpi, he investigated the power of magnets, and must at least have had his own views on the scientific method reinforced by Gilbert's lusty attacks on those who gained all their learning from books. He also made the acquaintance (through correspondence) of Johannes Kepler, an enthusiastic supporter of the Copernican idea that the Earth moves around the Sun. It was in a letter to Kepler written in May 1597 that Galileo first expressed in writing his own support for the Copernican cosmology. And alongside all this, Galileo studied literature and poetry, played the lute for pleasure, attended the theatre, and enjoyed an evening out drinking with his friends. He was 40 in 1604, and throughout his thirties he had lived life to the full, established a reputation for the practical application of science of immense value to the Venetian state, become a popular lecturer, and begun to gain a reputation as an anti-Aristotelian – something which, if anything, only enhanced his reputation in free-thinking Venice and Padua. And his early reputation as 'the wrangler' had been borne out, as he became known as a man well able to argue a case, often using biting wit and sarcasm (as well as impeccable scientific arguments) to reduce the case of his opponents in debate to ruin. His post at the university had not only been renewed, but with a salary increase. All in all, life was pretty good for the 40-year-old Galileo.

It was also in 1604 that Galileo first appeared in public as an astronomer. This was the result of the appearance in the sky of a 'new' star, or nova, in October that year – a star so bright that today it would be called a supernova. We now know that supernovas are, in fact, old, dim stars that suddenly flare into unusual brightness in one last burst of explosive activity at the end of their lives.

According to Aristotle, the celestial sphere beyond the Moon was perfect and unchanging. So any new heavenly phenomena, such as novas or comets, must exist in the region between the Earth and the Moon. Galileo studied the new star carefully, using surveying techniques (he was, after all, an expert military surveyor) to see if it showed any sign of movement relative to the other stars, but there was none. In a series of widely acclaimed public lectures, he argued that this object was indeed at a distance typical of the known stars, and that therefore, contrary to Aristotelian doctrine, the celestial sphere was not perfect and unchanging. This led to a fierce debate with the Professor of Philosophy in Padua, Cesare Cremonini, who defended the Aristotelian position (fierce, but amicable; there was no personal animosity between Cremonini and Galileo, who remained friends). It's interesting that although this public debate was important, Galileo was considerably less advanced than Gilbert in his astronomical thinking at this time, although he had by now read *De Magnete*. All of this activity enhanced Galileo's growing reputation, however, and helped to convince him of the value of the experimental method rather than abstract philosophy.

During the summer of 1605, Galileo visited Florence, where he had pressing business to attend to. Both his brothers-in-law were suing him for non-payment of the instalments on the dowries of his respective sisters, and he was only being kept afloat there by his friend

Sagredo, who had paid court fees and was stalling the legal process as best he could. In spite of his professional success and happy position in Padua, Galileo had always wanted to return to Tuscany, and in particular to get a Court appointment, which would free him from the time-consuming necessity to give lectures. Now, he was presented with a golden opportunity to promote his cause – the Grand Duchess, Christina, asked Galileo to give her son, Cosimo, tuition in the use of his military compass. With Galileo seen to be in high esteem in the Court, the litigation against him faded away, perhaps because Christina expressed disapproval, or perhaps because the litigants now reckoned that the money would be available to them soon if Galileo's star remained in the ascendant.

Until then, Galileo had only circulated manuscript copies of his instructions in the use of the compass, mainly because he wanted to restrict the knowledge as much as possible, so that he could still charge tuition fees for explaining its use. But in 1606 he published the manual in a printed book, in a strictly limited edition of 60 copies, dedicated to Prince Cosimo.

While Galileo continued to make overtures to the Tuscan Court, a political dispute flared up between Venice and Rome. A new pope, Paul V, had been elected in 1605, and as he tried to exert the authority of the Vatican in matters of taxation and border disputes about the sovereignty of the Doge of Venice, matters escalated to the point where, on 17 April 1606, the Doge was excommunicated. The Jesuits were expelled from Venice in retaliation, and war seemed inevitable; but eventually the crisis passed, and Venice retained its large measure of independence from Rome.

During this crisis, a leading figure on the Venetian side was Friar Paolo Sarpi, twelve years older than Galileo, who had befriended Galileo on his arrival in Padua and whom Galileo looked up to as 'my father and my master'. Sarpi incurred the hatred of the Jesuits, as did anyone associated with him; in October 1607, an assassination attempt left Sarpi desperately wounded, but he recovered; the would-be assassins escaped to Rome, where they were well treated by the Pope.

With all this going on around him, and his campaign for a post in Tuscany gaining pace with the ascent of Prince Cosimo to the throne (as Grand Duke Cosimo II), Galileo continued to develop his masterwork, a book on mechanics, inertia, and motion. He also studied magnetism, hydrostatics, and the strength of different materials, and it was at about this time that he realized, and proved, that the path of a projectile fired from a gun, or thrown through the air, is a parabola. Until this proof, it was still widely thought that if, for example, a cannon fired its ball horizontally, the ball would travel a certain distance in a straight line, and then fall straight down to the ground, rather than following a curved trajectory. Even those thinkers who had noticed that the actual trajectory followed by a cannonball was a curve did not know what *kind* of curve it was.

In 1609, Galileo was in his pomp: a leading figure in the Venetian state, 45 years old, secure in his post, still with financial worries, but working to complete what he knew would be a great book. Even if the views expressed in that book displeased the Pope, he would be safe to express them as long as he stayed where he was. But he still hankered after the freedom that an appointment as Court Mathematician in Tuscany would bring him – freedom from lecturing, financial security, and a return to the hills that he loved. In the summer of 1609, the opportunity to fulfil all those dreams fell into his lap.

The telescope had been invented by Hans Lippershey, in Holland, the previous year (actually, re-invented; as we mentioned earlier, Leonard Digges, in England, built telescopes before 1550). News of the Dutch device spread quickly across Europe, and Paolo Sarpi heard of it in a letter from one of Galileo's former pupils, a French nobleman then living in Paris, before the end of the year. It isn't clear whether the news didn't reach Galileo at that time, or whether he dismissed the story as fiction; perhaps Sarpi didn't take much notice and failed to pass the news on. But in July 1609 Galileo visited Venice, where he discussed the invention with Sarpi, and was shown the letter in which the telescope was described. At that point, we can imagine the dollar signs appearing in his eyes like a cartoon character with a get-rich-quick scheme. An instrument which could make distant objects visible would be of enormous value to a maritime power such as Venice.

While Galileo was in Venice, at the beginning of August news came that a stranger had arrived in Padua with one of the miraculous instruments. Galileo rushed home to find out more, only to discover that the stranger had already departed for Venice, where he intended to sell the telescope to the Doge. Aghast, and with very little to go on, Galileo began a frantic burst of experimentation, trying out lenses in tubes to make his own telescope. The design that he hit upon used one convex lens and one concave; unlike the Dutch version (with two convex lenses) this happens to give an upright image of the object being viewed. Within twenty-four hours, Galileo had a working telescope. On 4 August, a pre-arranged message was despatched to Sarpi, who had just been asked by the Venetian Senate to advise them about the telescope now being offered to them by the itinerant Dutchman.

Sarpi effectively froze the Dutchman out of the picture, giving Galileo time to build a beautiful instrument, with a magnifying power of 10 times, in a tooled leather case. He arrived back in Venice before the end of the month, and demonstrated his instrument to sensational effect, among other things giving the senators a clear view of galleys far out to sea, two hours before they became visible to the naked eye. In a final gesture of genius, showing his political nous, Galileo presented the telescope to the Doge as a free gift. In return, he was offered tenure in his post in Padua for life, at double his present salary.

Galileo accepted, but then discovered the snags. The increased salary was not to be paid until the following year, and tenure for life also meant teaching for life. A quick visit to Florence gave him an opportunity to demonstrate the telescope to Cosimo II, and by December 1609 Galileo had constructed a telescope with a magnifying power of 20 times, which he turned to the heavens and used to discover the four brightest moons of Jupiter early in 1610. He named them the Medician stars, in a further attempt to curry favour in Florence. With his powerful new telescope (about as powerful as a decent pair of modern binoculars), Galileo looked at the Milky Way, and found that it was made up of a myriad of individual stars. He looked at the Moon, and found that instead of being a perfectly smooth sphere it was pockmarked with craters and had mountains which he measured (correctly, from the length of their shadows) as 4 miles high. In March 1610, these amazing discoveries were published in his book *The Starry Messenger*, dedicated (of course) to Grand Duke Cosimo II de' Medici.

The book was such an extraordinary success that Galileo got only six copies of the first printing of 550. It has been described as the most important book of the seventeenth century, and only five years after its publication it had been translated into Chinese and published in China. The whole literate world knew of Galileo and his discoveries. The book was also a success in assisting Galileo's campaign in Florence. Having decided that he had no obligations in Venice or Padua because he had not yet started receiving the promised increase in salary, he campaigned shamelessly for a post in Tuscany, and in May 1610 his efforts were rewarded. He was to become Chief Mathematician of the University of Pisa and Philosopher and Mathematician to the Grand Duke for life, with a salary of 1,000 crowns a year, no teaching obligations whatsoever, and with the Grand Duke releasing him from the legal obligation to cover Michelangelo's remaining unpaid share of the two dowries, on which Galileo had more than covered his own original obligations. He was to take up the post in October.

The turmoil of the move and its implications completely altered Galileo's private life, as well as his professional prospects. His two daughters moved to Florence just before Galileo, initially staying with his mother; his son stayed in Padua with Marina Gamba until he was old enough to join Galileo. But Marina stayed in Padua, seemingly as a result of an entirely amicable split. The passion had died, and she didn't want to move, while Galileo didn't much mind leaving her behind. In Florence, Galileo soon made the friendship of a young nobleman in the mould of Sagredo – Filippo Salviati, twenty years younger than Galileo, who was wealthy, interested in science, and loved fine wine and good food. There could be no doubt that Galileo had fallen on his feet, although his pleasure was temporarily dampened by one of his recurring bouts of ill health early in 1611.

Before Galileo left Padua, he discovered that there was something strange about the appearance of Saturn – only after Galileo's death was this strange appearance explained as being due to a system of rings around the planet. Soon after he arrived in Florence, he discovered the phases of Venus, in which the appearance of the planet changes in the same way that the appearance of the Moon changes as it orbits the Earth. This was a killer blow to Aristotelian cosmology, since the only explanation for the phases is that Venus orbits around the Sun.

The desperation with which the Peripatetics tried to explain the new discoveries is highlighted by one bizarre suggestion, made by one of the philosophers in Florence. He said that the mountains and craters Galileo had observed on the Moon were real, but that all of these surface irregularities were encased in an invisible sphere of perfectly smooth, perfectly transparent crystal! Galileo's retort was that he would be happy to accept this possibility, provided he was allowed to claim that there were mountains of this crystal ten times higher than any mountains he had actually observed on the Moon.

Around this time, Galileo also observed sunspots, unaware that these had already been seen by other European astronomers with the aid of telescopes (they had long been known in China, from naked-eye observations when the Sun is low on the horizon and seen through haze). All of these discoveries flew in the face of Aristotelianism, and it was important to find out how the Church would react to the news. Once safely settled in Florence, there was one place Galileo just had to go to spread the word of all this – Rome.

Unlike Venice, Tuscany was very much involved with the politics of Rome, and the Medicis numbered cardinals and popes among their ranks. To put things in a broader perspective, this period was the height of the Counter-Reformation, the period roughly a hundred years long in which the Catholic Church was involved in a vigorous campaign against Protestantism. You might think that this would be a bad time to question the Aristotelian tradition that the Church had made its own. But in the spring of 1611, Galileo was fêted almost as much in Rome as he had been in Venice and Florence. He was careful not to proclaim the death of Aristotelianism, preferring to let the evidence speak for itself as he showed influential people, including cardinals, the wonders to be seen with the aid of a telescope. The Pope granted him an audience, and he was made a member of what is reckoned to be the first scientific society in the world, the Lincean Academy ('Academy of Lynxes'), founded in 1603. Largely thanks to Galileo's influence, this talking shop soon became an important body, the first independent group of scientific thinkers, unattached to any university and owing allegiance to no one. It was, incidentally, at one of the society's dinners during Galileo's visit that the term 'telescope' was coined.

Among the old friends Galileo met in Rome was Robert Bellarmine, now a cardinal (and member of the Inquisition) himself. And the records show one small cloud on the horizon associated with this visit. Probably because Galileo did cautiously mention to Bellarmine that his astronomical observations could be seen as supporting the Copernican cosmology, with the Earth moving around the Sun, Bellarmine wrote to the Inquisition in Venice to check whether Galileo had ever been formally implicated in teaching Copernican ideas. But there the matter rested – for the time being.

Back in Florence, Galileo continued his astronomical observations (he was particularly keen to find a way to use the changing positions of Jupiter's satellites as a kind of cosmic clock, so that navigators might determine the time accurately and therefore calculate their longitude, the key problem in navigation in the seventeenth century). He also studied hydrostatics, and wrote a little book about the subject which sold out twice in 1612. Even before his visit to Rome, he had begun to use the magnifying power of lenses as a form of microscope, magnifying the parts of insects to the wonder of his acquaintances. But the most important activity he became involved in at this time was the study of sunspots.

Galileo wrote a book on sunspots, published by the Lincean Academy in 1613. In a flowery preface, the Linceans asserted that Galileo had discovered sunspots. This led to a bitter row about priority with a German Jesuit, Christopher Scheiner, who also claimed priority for the discovery (ironically, though, telescopic observations of sunspots had already been carried out, and published, before either of them got in on the act, by the Dutchman Johann Fabricius, so the argument was pointless). Scheiner claimed that sunspots were tiny planets, orbiting close to the Sun; Galileo proved that they were blemishes on the face of the Sun, and that their motion was caused by the rotation of the Sun. In other words, the Sun itself was imperfect. Apart from the fact that this argument made another bitter enemy for Galileo in the Jesuit camp, the work on sunspots is especially interesting because it was in an appendix to the sunspot book that for the first and only time in print Galileo came out unambiguously as a supporter of

Copernican cosmology, citing the evidence of the movement of the 'Medician stars' around Jupiter and the way these moons are regularly eclipsed.

Although otherwise still extremely cautious about what he put down in print, Galileo began to speak out more openly in support of Copernican ideas, and to argue that the Bible should not be regarded always as the literal truth, but sometimes as metaphor. Complaints were made, in a general sort of way, to the Inquisition in Rome, but without any specific charges being raised against Galileo. No action was taken against him, and at the end of 1615 Galileo voluntarily visited Rome (with the official permission of his Grand Duke, and staying in the Tuscan Ambassador's residence while in Rome) to find out exactly where he stood, and to argue the cause of the new astronomy if the officials there would listen. This was a serious misjudgement, which brought matters to a head in an unwelcome fashion.

The expert theologians deliberated on the two key issues, and responded with an opinion backed by the full weight of the Catholic Church. It was officially decreed that the idea that the Sun is in the centre of the world was 'foolish and absurd . . . and formally heretical' while the idea that the Earth moves through space was decreed to be 'at the very least erroneous in faith'.

On 24 February 1616, the Pope instructed Cardinal Bellarmine (who was known to be sympathetic to Galileo) to notify Galileo of the decision. The Pope's instructions were quite specific – Bellarmine was to tell Galileo that he could not 'hold or defend' either of the notions that had been judged. If, and only if, Galileo objected to this, he was to be warned formally by the Inquisition, in the presence of a notary and witnesses, that he must not 'hold, defend, or *teach*' (our italics) these Copernican ideas.

The distinction was crucial. If he was allowed to teach Copernicanism, Galileo could use it as an example of what other people thought, even if he did not (at least, not officially) hold those views himself. At this point, events become cloudy.

There seems to have been a single meeting between Galileo and Bellarmine, with the representatives of the Inquisition, notary and witnesses all present. What seems to be an official minute of that meeting, but unsigned, records that Bellarmine issued the first set of instructions to Galileo, but that immediately, with no time for Galileo to react, the other representatives of the Inquisition issued the second warning, including the crucial reference to teaching.

The historian Stillman Drake has reconstructed the probable course of events, and argues persuasively that the deposition remained unsigned because Bellarmine (the senior official present, and the direct representative of the Pope in this matter) was furious that the Pope's explicit instructions had not been obeyed, and regarded only his initial warning to Galileo (who had not objected) as having any legal standing. Indeed, when rumours began to spread that Galileo had been punished in some way by the Inquisition, Bellarmine supplied him with a formal affidavit stating that this was not the case, and that Galileo had merely been informed of the general edict, which applied to all Catholics equally. The two documents – the unsigned 'minutes' of the meeting and Bellarmine's signed affidavit – were to prove crucial in Galileo's eventual trial for heresy.

Of course, that climax to the story is what we are now rushing towards, and we will say little more about Galileo's other work – his continuing studies of the moons of Jupiter, his investigations of comets, the slow progress with his great book, and more. But we cannot move on without mentioning one strange aspect of his private life.

In the spring of 1617, Galileo moved into a fine house known as Bellosguardo, on a hill to the west of Florence. The main reason for the move was that he wanted to be near his daughters, Virginia and Livia, who were entering the order of the Poor Clares, in a convent in nearby Arcetri. They had no choice – Galileo seems to have been determined that his daughters would spend their lives as nuns, and from a practical point of view, after his bad experiences with the dowries of his sisters, we can see why he was eager to ensure that he would neither have to support his daughters as unmarriageable illegitimate children, nor pay a substantial dowry to get them off his hands.

The fact that he could do this even though he loved them dearly enough to move to be near them looks odd to modern eyes, especially since the Poor Clares really were poor, and he was committing the girls, at the ages of 16 and 15, to a life of hard work, cold accommodation and inadequate food. But that is what he did, and in spite of everything stayed particularly close emotionally to the older girl, who took the name Maria Celeste; her sister became known as Arcangela.

In 1617, Galileo was 53 years old. As well as his longstanding illness, he now suffered from a severe hernia. He had achieved fame and (relative) fortune, and had been warned off by the Pope, if only in a gentle sort of way, from getting too closely associated with Copernicanism. A year later, the Thirty Years War (a Catholic–Protestant clash) began, and common sense would have ordained that he avoid annoying the Church further. But, again in 1618, three comets were seen, and this led Galileo back into public debate with the Jesuits, in the form of Orazio Grassi.

Grassi published a book about comets, to which Galileo replied in withering terms. We won't go into details, since both the Jesuits and Galileo were wrong, on this occasion, in their explanations of the phenomenon. But in his book *The Assayer*, published in 1623, Galileo also summed up his understanding of the scientific method. Sarcastically suggesting that his opponents seemed to think that 'philosophy is a book of fiction by some author, like *The Iliad*', he said that the book of the Universe

> cannot be understood unless one first learns to comprehend the language and to understand the alphabet in which it is composed. It is written in the language of mathematics, and its characters are triangles, circles, and other geometric figures, without which it is humanly impossible to understand a single word of it; without these, one wanders about in a dark labyrinth.

In other words, the Jesuits were dealing in fairy tales, while Galileo was dealing in facts.

Just as *The Assayer* was about to be published, the Pope died (this was Gregory XV, who had succeeded Paul V in 1621), and was replaced by Maffeo Barberini, who took the name Urban VIII. This seemed to be a stroke of luck for Galileo. The two had been on friendly terms when Barberini was a cardinal, and even better Galileo had tutored one of Barberini's nephews, Francesco, who received his doctorate in Pisa a few months before Urban VIII was elected. In an archetypal example of nepotism, Francesco Barberini became a cardinal himself almost as soon as his uncle was made Pope; another friend for Galileo in the Vatican. The change of pope seemed more than sufficient compensation for the death of the Grand Duke of

Tuscany, Cosimo II, in 1621; Cosimo had been succeeded by the 11-year-old Ferdinando II, with his mother Christina as regent, seriously weakening the voice of Tuscany in Italian politics, and therefore weakening Galileo's own political power base. In 1623, there was just time to ensure that the printed version of *The Assayer* was dedicated to the new Pope, and Galileo was delighted to hear that Urban VIII had enjoyed it so much that he had had extracts read aloud to him while dining.

In April 1624, Galileo visited Rome to pay his respects to the new Pope, and to Cardinal Barberini. He stayed until June, and was granted six audiences with the Pope, who gave him a gold medal and other honours, and wrote on his departure to the young Duke Ferdinando II, praising Galileo and his science. Galileo left under the clear impression that he had permission to publish his long-planned book comparing the Copernican and the Aristotelian world views, provided that he only argued the Copernican case hypothetically, and did not claim that science proved that the Earth moved.

In Galileo's own mind, of course, the case *was* proven. Ironically, though, what he regarded as his best evidence was completely mistaken. Perhaps naturally for an inhabitant of the Mediterranean region, Galileo likened the tides to the motion of water sloshing to and fro in a tub when the tub was moved. He was so sure that the tides proved the motion of the Earth that he planned to call his book 'Dialogue on the Tides' – and he was dissuaded from doing so because his friends suggested that the evidence from tides in support of Copernicus was so strong that mentioning the word in the title of the book would be sure to rouse the ire of the Church.

A much more important feature of Galileo's work, to modern eyes, was his discovery of the idea of inertia, the idea that Gilbert had come so close to with his image of the Earth rolling endlessly through space. By rolling balls down inclined planes and allowing them to roll up another plane, Galileo realized that in the absence of friction a ball would always roll up to the same height that it had rolled down from (that throwaway phrase 'in the absence of friction' conceals another great leap made by Galileo, who – as in the discussion of balls dropped from the Leaning Tower – was one of the first scientists to grasp and understand the idea of extrapolating from our imperfect experiments to some idealized world of pure science). But this was now nearly a quarter of a century after the publication of *De Magnete*, where Gilbert made such impressive use of idealized models of the world. How much Galileo was influenced by his reading of Gilbert's book we shall never know; but he certainly took the idea to new heights.

Galileo realized that in the rolling ball experiment if the second plane was made more shallow, the ball would roll further, until it got back to its original height. And if the second plane was actually horizontal, and friction could be ignored, the ball could never regain its lost height, and so it would roll forever. He inferred that motion, rather than just a state of rest, was a natural state of things. His one mistake was that because he knew the surface of the Earth is round, so that horizontal motion (motion towards the horizon) means following part of the curved surface of the Earth, he thought that this natural inertial motion must follow a circle – neatly explaining, it seemed, why the planets orbited the Sun continually without being pushed.

Over the next few years, Galileo worked only intermittently on the book, plagued by ill health, sometimes putting it to one side when the political climate seemed to shift against free-thinking, sometimes getting sidetracked by other work. He also had to fight a tedious legal battle with the University of Pisa, which had attempted to get out of the legal obligation imposed on them by Cosimo II to pay Galileo for doing nothing, even when he didn't live in Pisa. We have some sympathy with the university, although this is a good example of the kind of hassle that, surely, would not have arisen if the reigning Duke had been an adult; eventually, Galileo won the case. And, under repeated prodding by his friends (who must have begun to wonder if he would die without completing it), the book did progress. By the beginning of 1630 it was completed, and needed only the approval of the censors in Rome before it could be published.

Now, things began to get complicated. We'll smooth over the complications as much as possible, but the key thing to understand is the way Galileo's book, *Dialogue Concerning the Two Chief Systems of the World*, was constructed. As the title suggests, it was in the form of a conversation between two people, one supporting each of the world views. This was an old device, going back to the Ancient Greeks. Galileo added a third 'voice' in the book, a supposedly independent observer who listened in on the conversation and raised points for debate. The character who presented Galileo's views (arguing the Copernican case) was called Filippo Salviati, after Galileo's old friend, who had died in 1614. The independent observer was named Giovanfrancesco Sagredo, after his other friend, who had died in 1620. The supporter of Aristotle (or rather, strictly speaking, of the Earth-centred cosmology of Ptolemy) was called Simplicio, after an Ancient Greek who had written a commentary on Aristotle's work. This could be presented (tongue firmly in cheek) as an innocuous choice of a name borrowed from a genuine supporter of Aristotle; but Galileo undoubtedly intended to imply that anyone who supported Aristotle was a simpleton, and this choice of name was to be a major factor in getting him into hot episcopal water.

The book slowly made progress towards publication through 1630, with the censor demanding minor changes, but taking no exception to the general tone of the book. A preface was required, spelling out the hypothetical nature of the ideas discussed in the book, and some words had to be added at the end, making it clear that Aristotle was the approved choice of the Church. The point the Church wanted to get across in allowing the book to be published was that they were not scientific ignoramuses, lagging behind the rest of Europe, but were well aware of the scientific debate, and had made their decision on religious grounds, which brooked no argument, scientific or otherwise. Originally, the book was to have been published in Rome by the Linceans, but the death of Prince Frederico Cesi, the chief 'lynx', threw the affairs of the Academy into turmoil,[10] and permission was obtained to transfer the printing to Florence. An outbreak of plague made travel between Florence and Rome difficult, and delayed matters further. At last, printing began in June 1631, and the finished copies of the book (an edition of a thousand copies) went on sale in Florence in March 1632, with some copies being sent to Rome as fast as the plague permitted.

Cardinal Barberini wrote to Galileo expressing his delight with the book; but others were less pleased. Galileo's old Jesuit enemies were roused by, among other things, the repetition in the *Dialogue* of his

10. It faded out of existence altogether not long after.

claim to have discovered sunspots first. Somebody noticed that the preface required by the censor had been set in a different type from the rest of the book, effectively distancing Galileo from it. And the Pope's own words, stating the truth of the Aristotelian world view, had been placed in the mouth of Simplicio. Could it be that Galileo was saying that the Pope was a fool?

In fairness, since Simplicio is the only character in the book who supports Aristotle, the words insisted on by the censor could hardly have been spoken by either of the other characters. But the choice of name that must have seemed such a great joke to Galileo when he started writing the book turned out not to have been a good move. And then, with the ground prepared by having pointed all this out to the Pope, some unknown person with a grudge against Galileo dug out of the files the unsigned minute from 1616, forbidding Galileo to 'hold, defend, or *teach*' the Copernican world view. On the evidence he had before him, it seemed to Urban VIII that Galileo had deliberately flouted the orders of one of his predecessors.

All of this roused the Pope to a fury. Galileo was summoned to Rome to stand trial for heresy – for writing a book which had been passed by the official censor! Pleading old age and (genuine) infirmity, he postponed the evil moment as long as possible, but on 13 April 1633, when Galileo was in his 70th year, the infamous trial began.

Galileo was totally unfazed. He trumped the production of the unsigned minute by producing the signed affidavit from Bellarmine (who had died in 1621), which required him only to refrain from holding or defending Copernican ideas, not from teaching them hypothetically. He claimed that since Paul V and Bellarmine had put him right in 1616, he had indeed stopped believing such nonsense, and only taught it out of scientific interest. By any conventional legal test, Galileo had won – his documentary evidence beat that of the prosecution into a cocked hat, quite apart from the fact that his book literally had the official seal of approval. But nobody could beat the Inquisition. Once the wheels were set in motion, somebody had to take the rap, not least since to bring a false charge of heresy was as great a crime as heresy itself.

With Cardinal Barberini acting as intermediary, a long process of what would now be called plea bargaining followed. Galileo understood that if he admitted wrongdoing, that in his enthusiasm to present sound arguments he had gone too far in his book, he would get a light sentence. But he didn't think he had done wrong, and at first refused to confess, not really believing that the alternative was torture. He made the confession grudgingly (after some heavy persuasion by Barberini, who realized only too well what might happen to his old friend, and showed him the instruments that would be used to extract a confession anyway if he did not give in gracefully), and was stunned when, on 22 June, the formal sentence of life imprisonment was read out to him at a ceremony in front of the assembled cardinals of the Inquisition. Even then, though, three out of the ten cardinals (including Barberini) refused to sign the sentence; he went down only on a majority verdict.

It sounds harsh, but remember that in those days the maximum sentence he could have got was the rack followed by being burned at the stake. By some criteria, life imprisonment *is* a soft option. And it was Cardinal Barberini who ensured that the punishment started out in the grounds of the Tuscan Embassy in Rome, then shifted to the custody of the Archbishop of Siena (another Galileo sympathizer) and ended up in Galileo's own home, now in Arcetri, as nothing more than house arrest, from the beginning of 1634. Sadly, on 2 April that year, his favourite daughter, Sister Maria Celeste, died.

In his enforced 'retirement', Galileo overcame his many disappointments, and turned back to his work, completing his greatest book, *Two New Sciences*,[11] which was smuggled out of Italy in manuscript and published in Leyden in 1638 by Louis Elzevir. This book summed up all of Galileo's work, on mechanics, inertia, pendulums, the strength of bodies, and the scientific method. It was of enormous influence in the decades that followed – and, like all of Galileo's works, assured of an enthusiastic readership throughout Protestant Europe precisely because of Galileo's conflict with the Catholic establishment. It is no coincidence that after the 1630s Italian science went into decline, while dramatic new developments occurred to the north and west, especially in Britain – among Galileo's visitors in his final

11. In full, *Discourses and Mathematical Demonstrations Concerning Two New Sciences.*

years were Thomas Hobbes, who brought news of an English translation of the *Dialogue*, and later John Milton, on a tour of Europe at the age of 29.

By the time *Two New Sciences* appeared in print, Galileo had gone blind. Even after losing his sight, he invented an escapement for a pendulum clock, which he described to his son Vincenzio; the clock was built after Galileo died, but the spread of similar clocks across Europe followed the independent work of Christiaan Huygens. In Galileo's last years, from late in 1638, Vincenzio Viviani joined him to act as his scribe and assistant. Galileo remained mentally active to the end, though growing physically more frail, and died peacefully in his sleep, on the night of 8/9 January 1642, a few weeks short of his 78th birthday.

There is no evidence that he uttered the words '*eppur, si muove*' (yet it does move) as he was led from the Inquisition after formal sentence was passed. If he had, he surely would not have lived for the best part of a further nine years.[12]

It is impossible to overstate the importance of Galileo in establishing the scientific investigation of the world. His specific scientific achievements were impressive enough. He built with his own hands the telescopes (the best of his day) that he used to study the skies, and through those observations he found compelling evidence that the Earth is not at the centre of the Universe. He was the first person to appreciate the role of forces in determining the way things move – to the Peripatetics, some things just had a natural tendency to rise, while others just had a natural tendency to fall, but Galileo said that all motion except his version of inertial motion must be due to a force.

The idea of inertia that he developed was not quite right, but an impressive step forward from Aristotelian ideas. This idea was taken up by René Descartes, who realized that it actually applied to motion in a straight line, not to circular motion; from Descartes, the idea of inertia was picked up, as we shall see, by Isaac Newton (via Robert

12. In 1992, a commission under Cardinal Paul Poupard found that Galileo had been 'more perceptive' in his interpretation of the Bible than his prosecutors, and that, as he suggested, the Bible should not be regarded as always telling the literal truth, but sometimes as metaphor. The Pope, John Paul II, formally pardoned Galileo on 31 October 1992, some 350 years after he had died. *Eppur*, indeed, *si muove*.

Hooke), and became the basis of his first law of motion. And since the real 'natural tendency' of an object such as a planet is to move in a straight line unless acted upon by an external force, that in turn was a key insight in developing the law of gravity, explaining why the planets are held in orbit around the Sun.

To put all this into perspective, as late as the 1620s the philosophers were still arguing about what would happen to a heavy object dropped from the top of the mast of a moving ship. Notice that they did not drop things from the masts of moving ships to find out what happened; they just argued the issue to arrive at the answer by pure reason. Some said that the object would be left behind by the motion of the ship; others that it would move forward with the motion of the ship and fall at the foot of the mast. Of course, the second idea is correct, as Galileo was careful to find out by talking to mariners. The falling object preserves the forward motion it started out with (Gilbert had realized this long before, but that didn't stop the philosophical arguments). The definitive version of this specific experiment was carried out as late as 1640, only two years before Galileo died, when the Frenchman Pierre Gassendi borrowed a galley from the French navy and dropped a series of balls from the mast to the deck below while it was being rowed flat out across the smooth Mediterranean (a grim modern example of this conservation of momentum can be seen in any film footage of a string of bombs falling from an aircraft).

Gassendi was strongly influenced by Galileo's writings, and this highlights the real revolution that Galileo brought to the study of the world. He promoted the Gilbertian idea of scientific experiments, and the whole business of testing hypotheses by getting your hands dirty investigating the real world, instead of strolling about discussing it in philosophical terms. This experimental approach was enthusiastically espoused by the Royal Society, and firmly established, well before the end of the seventeenth century, as *the* scientific method. Ironically, though, the Royal specifically picked up the idea neither from Gilbert nor Galileo, but from a philosopher of science, a contemporary of Galileo but a man who, as we shall see in the next chapter, devoted a lot of thought to the scientific method while scarcely ever getting his hands dirty by carrying out experiments. A classic example of 'do as I tell you, not as I do'.

# 2

# The Philosopher Scientist

The trial of Galileo provides a convenient marker for the shift of the centre of scientific gravity away from Italy, before too long to settle in England. Of course, significant investigation of the world around us continued to be carried out even in Italy after the trial of Galileo; and there was plenty of scientific investigation of the world going on outside England in the seventeenth century. But somewhere had to take the lead, and that somewhere was England. Curiously, though, when the scientific revolution took place in the middle of the seventeenth century, and the Royal Society was established as the leading centre of scientific investigation of the world, the people responsible did not consciously acknowledge their debt to William Gilbert, or even to Galileo. Rather, they saw themselves as putting into practice the abstract ideas of a philosopher scientist who had himself scarcely carried out an experiment in his life, but who described an idealized way in which science ought to be done. 'Baconian' science as applied in the early days of the Royal owed as much (or as little) to Bacon's philosophy as the various forms of 'Marxism' practised in the twentieth century owed to the teachings of Karl Marx. It is a moot point whether what the founders of the Royal Society did was actually what Francis Bacon described in his philosophy; but he became a symbol of what they were trying to achieve, and the end result was the same – things got done.

Bacon was born in 1561 and died in 1626, so there was considerable overlap between his life and that of William Gilbert, and he was certainly aware of *De Magnete*, although there is no evidence that the two ever met. Like the Gilberts, Bacon's family provides an example of the rise of the new elite in Tudor England, after the Reformation

*Francis Bacon*

carried out by Henry VIII. Francis Bacon's grandfather was a yeoman, an East Anglian sheep-reeve, who may also have been bailiff to the monks at the abbey of Bury St Edmunds. He had just enough money and social standing to ensure that his son, Nicholas, was able (with the aid of a scholarship) to go up to Bene't College (since renamed Corpus Christi) in Cambridge in 1523, at the age of 13. From there, Nicholas went on to a career in law, entering Gray's Inn in 1532, becoming a gentleman, and rising to become Lord Keeper of the Great Seal of England on the accession of Elizabeth I in 1558, by which time he had become Sir Nicholas. This had previously been merely an honorary position, but in fact Elizabeth had appointed Sir Nicholas as her First Minister, replacing the old Lord Chancellor, Archbishop Heath, who had refused to follow her instructions, but without giving him the same title. This was a significant moment in English political history – the transfer of the role of First Minister from the Church to a layman.

The place of the Bacon family in Tudor society had been secured by not one but two judicious marriages. Nicholas Bacon's first wife was Jane Fernley, the daughter of a wealthy Suffolk merchant, who gave him seven children before she died in 1552. Her sister Anne married Sir Thomas Gresham, a merchant banker based in London who, among other things, was the financial agent for the Queen – a useful family connection. When Gresham eventually died without a male heir, he left his entire fortune to found Gresham College, which we shall hear more of in due course. But that is getting ahead of our story.

When Jane Bacon died, her six surviving children were all under the age of 12, and Sir Nicholas remarried within a matter of weeks, ostensibly to ensure the continued smooth running of the household. There is circumstantial evidence that his new bride, Anne Cook, had already been the subject of his attentions before his first wife died, and they were certainly devoted to each other. Nevertheless, the astute Sir Nicholas had ensured that his second marriage cemented his place in society. Anne was one of the five daughters (there were also four sons) of Sir Anthony Cooke, who had been tutor to Prince Edward, the son of Henry VIII and later King Edward VI, and ensured a good education for all his daughters. He lost his influence at Court (choosing voluntary exile) during the reign of Edward's sister Mary, but returned to favour with the accession of Elizabeth. His daughters all married men who became leading figures in the Elizabethan establishment, and would provide Francis Bacon with a network of influential uncles; the most important of these was William Cecil, who married Anne's sister Mildred, and became, as Lord Burghley, Elizabeth I's Principal Secretary of State.

Francis was a product of this second marriage of Sir Nicholas, together with an older brother, Anthony, born in 1558. Francis himself was born on 22 January 1561.[1] The boys were largely brought up in the countryside, at the manor of Gorhambury, near St Albans, which their father had purchased in 1557. As well as Anthony, Francis had three half-brothers from his father's first marriage, and as his wealth grew[2] Sir Nicholas bought more land which was settled on each of his sons in turn to ensure their future financial security. He had got as far as the fourth son, Anthony, when he died unexpectedly in 1579, leaving the 18-year-old Francis almost penniless, but accustomed to moving in the highest circles and with great ambitions. It was a combination that would cause him difficulties for the rest of his life.

The early education of Anthony and Francis Bacon had been largely the responsibility of their mother, who was both well educated and adhered to the strict Protestant religion. Under her influence, and the influence of tutors chosen by her, they became god-fearing and reasonably diligent students, unlike their three elder half-brothers. Francis also spent some time as a child in the household of his uncle, William Cecil (Lord Burghley), where he was educated alongside William's son, Robert, two years younger than Francis, who later became Lord Salisbury. There is a story that when he was a boy of 12

1. He would have given his birthdate as 22 January 1560, because at that time the New Year officially began on 25 March. We have regularized all dates to conform with the modern calendar, and also made similar allowance for the fact that the calendar in mainland Europe was ten days ahead of the one in England (so 22 January in England was 1 February in France).

2. His income as Lord Keeper was about £2,600 in 1560, and rose to £4,000 the year before he died. It is impossible to give accurate modern equivalents for incomes and prices in the sixteenth century, but as a rough rule of thumb they should be multiplied by at least 500.

Francis was introduced to the Queen, and when asked his age charmed her by saying that he was 'two years younger than Her Majesty's happy reign'. Although the three sons of Sir Nicholas Bacon's first marriage (Nicholas, Nathaniel, and Edward) all attended Cambridge University, they did so only briefly, with Edward being the only one to stay there more than a year. By contrast, Anthony and Francis went up to Trinity College together, in 1573, and lasted almost a full three years;[3] at 12, Francis was young to be attending university even for those times, but he had always been close to his brother, and the two boys were in good hands during their time at Trinity, where they were the personal responsibility of the Master, John Whitgift, who would later be Archbishop of Canterbury.

The education the brothers received was the Classical grounding in Greek, Latin, and logic, with such science (or natural philosophy) as there was devoted to the works of Aristotle. Much later, in his *Life of Bacon*, William Rawley, who was Bacon's chaplain, would describe how even at this precocious age Francis immediately 'fell into the dislike of the philosophy of Aristotle' because of the worthlessness of his method, which was 'only strong for disputations and contentions, but barren of the production of works for the benefit of the life of man'. The story should perhaps be taken with a pinch of salt; it is more likely that in old age Bacon recounted the tale which compressed his first encounter with Aristotle at the age of 12 or 13 with his later reasoned dislike of the Aristotelian method. But it is still significant that he should choose these words to express his dislike of Aristotle – because his method was 'barren of the production of works for the benefit of the life of man'. However old (or young) he was when he came to that conclusion, Bacon's belief that learning – what we would now call science – was useless unless applied to the service of humankind underpinned what became his own philosophy of science.

On leaving Cambridge, Francis continued to follow in his father's footsteps by being admitted to Gray's Inn in 1575. But his father's position and wealth as Lord Keeper gave him wider opportunities, and in those changing times it was already becoming not only desirable

3. With two interruptions when the boys returned home during outbreaks of the plague.

but necessary for any prospective high-flier to broaden his education on the continent. A convenient way for a suitably connected young man to gain such experience was to travel in the entourage of an ambassador – and no ambassador of Queen Elizabeth could refuse the honour of taking the Lord Keeper's son under his wing. So in 1576 Francis set off for Paris in the care of Sir Amias Paulet, the Queen's Ambassador to the French King, Henri III. As his later recollections make clear, it was one of the highlights of the 15-year-old boy's life that as a member of the Ambassador's party he was personally acknowledged by the Queen, and kissed her hand, on setting out on the journey.

In France, where the party reached Paris on 3 October 1576, Bacon lived at first as a member of the Ambassador's household, learning the ropes of diplomacy and statesmanship, and copying from Paulet what became his lifelong habit of working steadily at a given task and staying at it late until it was finished, rather than let the business hang over to the next day. His formal education under different tutors continued, but he also indulged in a little light spying (as almost everybody in Court circles did in those days), reporting back to his friend Thomas Bodley, who later donated his famous book collection to found the Bodleian Library in Oxford. This involved nothing so crass as stealing secret documents or other James Bond-like activities, but rather becoming deeply familiar with the complexities of the political situation in France (where Henri III was involved in a power struggle with his Protestant (Huguenot) rival, Henri de Navarre) and being paid to report back to Bodley the background to the various changing alliances and Court intrigues associated with this struggle. In this cloak-and-dagger atmosphere, codes and ciphers were essential for sensitive messages sent back to England, and Bacon also developed a lasting interest in these, including his own version of the binary code, in which all the letters of the alphabet were represented by combinations of *a* and *b*, so that *A* is *aaaaa*, *B* is *aaaab*, and so on down to Z (written as *babbb*, since at that time no written distinction was made between *i* and *j* or between *u* and *v*, so there were only twenty-four letters in the written alphabet). Of course, this is far too transparent to be a useful cipher for state secrets; but it is essentially the same binary code used to store and manipulate information (as strings of 1s and 0s, rather than *a*s and *b*s) in computers today.

It is a sign of how much promise Bacon showed that in the autumn of 1577 he was chosen by Paulet to take both verbal and written messages back to London, providing an opportunity for Bacon to be seen in the Queen's service, improving his contacts at Court and moving him another small step closer to the corridors of power. This was a real indication that Bacon had made his mark in Paulet's entourage; the verbal messages, in particular, would be likely to include information too sensitive to be trusted to paper even in code, bringing the bearer into direct, personal contact with the high-placed people the messages were intended for.

Not long after his return to France, at the beginning of 1578, Bacon moved out of Paulet's household into the home of a civil lawyer, where he had the opportunity both to study law and to improve his grasp of the French language. That wasn't all he studied. Paris was alive with intellectual debate at the time, particularly among men (women were almost entirely excluded) trained in the law and active in public service – just the circles that Bacon moved in. Regular gatherings of such learned men were held in an informal fashion to discuss philosophical issues and the material that would now be regarded as falling within the province of science, and there was widespread criticism of the old philosophy of Aristotle. Almost certainly, it was during his time in Paris that Bacon actually developed the anti-Aristotelian views that he later described as a product of his time in Cambridge; if he really had made up his mind about Aristotle already, he certainly found plenty of scope to reinforce and develop those views in Paris. It is indisputable that he held his own in this company. In 1578, when Bacon was 17, he had his miniature painted by Nicholas Hilliard. Around the image of Bacon's face Hilliard wrote a Latin inscription which translates as 'if only one could paint his mind'.

But on the brink of a glittering career in public service and the law, Bacon's immediate hopes were dashed in February 1579, when his father died at the age of 69, taken suddenly by a chill, a victim of severe winter weather. In March, just two months after his 18th birthday, Francis returned to England – again acting as a messenger for Paulet – to learn that he was almost penniless, and would have to find a way to make a living before he could achieve his ambitions to become a statesman.

For the rest of Queen Elizabeth's long reign, Bacon flirted with financial and political disaster, kept afloat by loans and always anticipating political preferment that never came. His problems were exacerbated by his extravagant lifestyle, taste for fine clothes, and an unconventional household in which he lived as a bachelor with an all-male retinue of servants who treated his property as their own. There was a well-documented occasion,[4] much later, when Bacon had become Lord Chancellor and at last received a large income, when a visitor who was left alone in a room while Bacon was temporarily absent watched in amazement as first one servant then a second came into the room, casually opened a chest where Bacon stored his ready cash and filled his pockets before wandering off. When Bacon returned and the visitor reported what had happened, he simply shook his head and said, 'Sir, I cannot help myself.' He was undoubtedly a homosexual, which is relevant to our story only in that it seems to have been linked to his extravagance, as confirmed by a letter from Lady Anne Bacon to her elder son Anthony, written in 1593, in which she says:

> Surely though I pity your brother, yet so long as he pitieth not himself but keepeth that bloody Percy, as I told him, yea as coach companion and bed companion, – a proud profane costly fellow, whose being about him I verily fear the Lord God doth mislike and doth less bless your brother in credit and otherwise in his health, – surely I am utterly discouraged . . .

The complications of Bacon's private and public life are, alas, outside the scope of our present book, but have been dealt with at length in the excellent biography *Hostage to Fortune*, by Lisa Jardine and Alan Stewart. The immediate relevant development, following the death of his father, was that Francis took up residence at Gray's Inn, in the family chambers with his brother Anthony. All of the Bacon brothers had been admitted to Gray's Inn, but in the case of the older boys (and initially Francis as well) this was not with any real intention of studying (let alone practising) the law. They entered under a rule

4. See Jardine and Stewart.

allowing admission to the sons of judges and lords, without any requirement to keep proper terms or study the law seriously – which is why Francis had been able to head off to Paris so soon after his admission to the Inn in 1576. Like many other young men in this fortunate position, the elder Bacon sons treated the Inn as a convenient base in London, and bothered with the law only enough to be a help to a country gentleman in his dealings with the world. From their parents' point of view, the arrangement meant that the young men were at least living in respectable quarters, and some kind of watch could be kept on their behaviour. But not much – the Inns were located conveniently for the theatres, brothels, and gambling houses that proliferated just outside the City of London proper, and their younger residents had a reputation not unlike that of the young men who worked in the City in the 1980s boom.

Francis was soon left alone at Gray's Inn, determined to avoid these temptations and make himself into a proper lawyer. Anthony had to wait until he was 24 before he could inherit his share of their father's estate, and chose to spend the time travelling on the continent. Francis continued his studies, plagued by the poor health which would affect him throughout his life, but emphasizing in correspondence to Burghley and others that he did so solely as a means to make a living, while what he really wanted was to follow in his father's footsteps in the service of the Queen. One of the first signs of the value of his family connections came in January 1581, when (still not quite 20 years old!) he became Member of Parliament for Bossiney, in Cornwall, in the Parliament which sat from 16 January to 18 March. In those days, Parliament only sat when it was summoned by the Queen, who ruled for most of the time with the aid of advisers such as the Lord Keeper and the Lord Chancellor. She needed Parliamentary approval, though, if she wanted to raise money from taxes, or if laws had to be introduced or amended. The snag was that, once a Parliament had been called, it might refuse her demand for funds (a 'subsidy'), or choose to pass laws or act in other ways that the Queen did not approve of (something that would before too long cause problems for her Stuart successors), so there was often some reluctance on the part of the sovereign to call Parliament at all.

Francis owed his place in the Parliament of 1581 (where, probably wisely, he made no contribution at all) to the death in 1577 of Sir Robert Doyly, the previous Member for Bossiney, who was the husband of Francis's half-sister Elizabeth. The choice of who represented a constituency in those days was entirely a matter of patronage – nothing so crass as a democratic vote – and the power of choosing the Member for Bossiney lay in the hands of the Earl of Bedford, who just happened to be Francis's godfather. Apart from giving Francis a brief taste of how Parliament worked, though, the experience was less significant than him being called to the Bar, on 27 June 1583. Although the rules specified that a further five years should elapse before he could be raised to the bench at the Inn, signifying that he was allowed to plead cases at the courts, Bacon was able to pull enough strings to achieve this status (as a 'bencher') in 1586, when he was 25, although not without ruffling a few feathers. This was to be a recurring theme for the next twenty years. His only hope of advancement was to pull strings and promote his own interests at every opportunity; but this gave an impression of desperation, and all too often was counter-productive, not least in his relationship with the most important person in the land, the Queen.

The one place Bacon could make a name for himself and attract the right kind of attention was Parliament, where his career really began in 1584, when he sat as Member for Weymouth and Melcombe Regis.[5] For the first time, he served as a member of committees, the groups of individuals assigned to investigate particular issues before reporting back on them to the House of Commons. And he made his maiden speech in the House, in which he unwisely drew attention, in passing, to the great favours his father had received from the Queen, contrasting this with his own position as the impoverished fifth son of the former Lord Keeper. This made a distinctly bad impression, but Bacon was soon to become a respected and consummate Parliamentarian, who served in every subsequent Parliament of Queen Elizabeth's reign, called upon for a great deal of committee work and known for his honesty (to the point of tactlessness) and capacity to see clearly to the

5. We won't concern ourselves here with the reasons why Bacon sat for different constituencies in different Parliaments, except to say that it was, as ever, all a matter of patronage.

heart of an issue. In 1586, he sat as Member for Taunton, and spoke in support of the execution of Mary Queen of Scots, then served on the committee that drew up a petition for her execution. That such a relatively junior Member should be chosen to represent the 'party line' in this matter was a clear sign of approval in high places; he began to get a little legal work on behalf of the ruling councils, and was even consulted on some legal matters by the Privy Council itself.

In 1588 (the year of the Spanish Armada), clearly as a result of more string-pulling (there is some evidence that the Queen herself may have given him the nod), Bacon became a Reader of Gray's Inn, an important step up the hierarchy at an unusually early age. But this still didn't solve his financial problems, and about this time Bacon began to pin his hopes on the influence of a new associate, the young Earl of Essex, who at the age of 24 had already distinguished himself as a soldier, and had been appointed Master of the Horse by the Queen in December 1587; the significance of the post is that it gave him the right of personal attendance on the Queen. Essex had been under the guardianship of Burghley as a boy, so he was very much part of the circle in which the Bacons moved, and both Anthony and Francis Bacon soon entered his service (Anthony eventually returned to England after ten years on the continent). Essex, a handsome and charming young man, famously became the favourite of the ageing Queen Elizabeth after the death of her old favourite, the Earl of Leicester, in September 1588; but even this did little to improve Francis Bacon's financial position. The only concrete reward for his increasing prominence in legal and Parliamentary circles and his constant attempts to curry favour was that in November 1589 he was granted the right of reversion to a senior legal post, that of Clerk of the Counsel in the Star Chamber. This meant that he could have the post, worth £1,600 a year, when it next became vacant; the snag was that the incumbent, who had no intention of relinquishing the post in his own lifetime, lived for another twenty years, with the tantalizing prospect of a decent income constantly dangling before Bacon for all that time.

By 1592, Bacon began to wonder if he would ever be able to achieve the high political office for which he craved. Not long after he turned

31, in January that year, he wrote one of his many letters to Burghley seeking in vain for some practical help in achieving these aims; but he also muses that 'I wax now somewhat ancient, one and thirty years is a great deal of sand in the hour-glass', before going on to give us one of the first hints of just how wide-ranging his interests were:

I confess that I have as vast contemplative ends, as I have moderate civil ends: for I have taken all knowledge to be my province; and if I could purge it of two sorts of rovers,[6] whereof the one with frivolous disputations, confutations, and verbosities, the other with blind experiments and auricular traditions[7] and impostures, hath committed so many spoils, I hope I should bring in industrious observations, grounded conclusions, and profitable inventions and discoveries; the best state of my province. This, whether it be curiosity, or vain glory, or nature, or (if one take it favourably) *philanthropia*, is so fixed in my mind as it cannot be removed.

Around the same time, Lady Anne Bacon writes in a letter to Anthony that:

I verily think your brother's weak stomach to digest hath been much caused and confirmed by untimely [that is, late] going to bed, and then musing *nescio quid* [on some nonsense or other] when he should sleep.

And later in 1592, when Essex organized an entertainment (known in those days as a 'device') to mark the Queen's birthday, Bacon contributed an item titled *Mr Bacon in Praise of Learning* in which he gives a clear idea of where those musings have led him:

Are we the richer by one poor invention by reason of all the learning that hath been these many hundred years? The industry of artificers maketh some small improvement of things invented; and chance sometimes in experimenting maketh us to stumble upon somewhat which is new; but all the disputation of the learned never brought to light one effect of nature before unknown.

All the philosophy of nature which is now received is either the philosophy of the Grecians or that other of the Alchemists. That of the Grecians hath the foundation in words, in ostentation, in confutation, in sects, in schools, in

6. The 'rovers' are, respectively, the Ancient Greek philosophers and the Alchemists.
7. That is, mysticism concerning gold, in alchemy.

disputations. That of the Alchemists hath the foundation in imposture, in auricular traditions, and obscurity. The one never faileth to multiply words, and the other ever faileth to multiply gold.

. . . we need [instead] the happy match between the mind of man and the nature of things. And what the posterity and issue of so honourable a match may be, it is not hard to consider. Printing, a gross invention; artillery, a thing that lay not far out of the way; the [compass] needle, a thing partly known before; what a change have these three made in the world in these times; the one in the state of learning, the other in the state of war, the third in the state of treasure, commodities and navigation. And those, I say, were but stumbled upon and lighted upon by chance. Therefore, no doubt, the sovereignty of man lieth hid in knowledge; wherein many things are reserved, which kings with their treasure cannot buy, nor with their force command; their spials [spies] and intelligencers can give no news of them, their seamen and discoverers cannot sail where they grow. Now we govern nature in opinions, but we are thrall unto her in necessity; but if we would be led by her in invention, we should command her in action.

It is from this time on that there is an increasing amount of Bacon's written material, both published and unpublished, which provides insight into his thinking about the scientific method, which preoccupied him throughout the long years when he was without an official post. Not that this distracted him from his efforts to secure such a post. Unfortunately for the immediate outcome of those efforts, though, in the early 1590s Bacon's honesty and expertise as a Parliamentarian led him into a direct conflict with the Queen.

In the Parliament of 1593, Bacon sat as the Member for Middlesex. Parliament met on 19 February, and a little later Essex was sworn in as a member of the Privy Council, the sovereign's innermost circle of advisers. If Bacon had been merely a self-seeking politician who cared only for his own advancement, he was now in the ideal position to speak up in the Commons in favour of whatever the Queen wanted, and to rely on Essex to recommend him for whatever plum job next became available. But he was not, and the issue which Parliament had been called to deal with brought the Commons into head-on conflict with the sovereign and the Lords.

The problem was money. Between 1558 (the year of the accession) and 1584, the Queen had been granted six subsidies, more or less evenly spaced through the period. Since two years was the usual time allotted to collect the subsidy, this gave the taxpayers a reasonable breathing space between subsidies. But her expenses had recently increased (largely, but not entirely, because of the threat of war with Spain, and the need to provide military aid to the Protestant opponents of Spain in the Netherlands), and at the end of the 1580s, in the run-up to the Parliament of 1593, she had been granted three subsidies in the space of just four years, each bigger than anything awarded to her predecessors, including a double subsidy (twice the usual tax) in 1589, which the country had only just finished paying. Now, she was asking for more.

The convention was that the Commons (which represented the taxpaying classes) would listen to the request, taking note of special needs such as the danger to the country from Spain, and set the level of the subsidy. The appropriate committee (on which Bacon sat) recommended a repeat of the double subsidy of 1589, as an extraordinary measure, which should not establish a precedent. Now, the Lords stepped in, sending Burghley to inform representatives of the Commons that a double subsidy was not enough, and that nothing less than a triple subsidy, paid at twice the usual rate (so that the whole thing would be paid in just three years, not six), would do. Along with some other Members of the House of Commons, Bacon saw this as an intolerable interference with their affairs. He spoke in the House in support of the payment of the subsidy, but not on the basis of a joint action by the Lords and Commons:

> For the custom and privilege of this House hath always been first to make offer of the subsidy from hence unto the Upper House. And reason it is that we should stand upon our privilege. Seeing the [tax] burden resteth upon us as the greater number, no reason the thanks should be theirs. And in joining with them in this motion we shall derogate from ourselves; for the thanks will be theirs and the blame ours, they being the first movers. Wherefore I wish that in this action we should proceed, as heretofore we have done, apart by ourselves, and not joining with their Lordships.

After further (sometimes angry) debate over the next few days, a joint meeting of the Lords and Commons took place, after which the

Commons proposed a triple subsidy to be paid over four years. While agreeing to the subsidy, Bacon spoke passionately about the need to spread it over at least six years, for the very sound reason that levying such a swingeing tax would cause widespread difficulty, and that there were better ways to find the money:

For impossibility, the poor men's rent is such as they are not able to yield it, and the general commonalty is not able to pay so much upon the present. The gentlemen must sell their plate and the farmers their brass pots ere this will be paid. And as for us, we are here to search the wounds of the realm and not to skin them over; wherefore we are not to persuade ourselves of their wealth more than it is.

The danger is this: we shall thus breed discontentment in the people. And in a cause of jeopardy, her Majesty's safety must consist more in the love of her people than in their wealth. And therefore we should beware not to give them cause of discontentment. In granting these subsidies thus we run into two perils. The first is that in putting two payments into one year, we make it a double subsidy; for it maketh 4s. in the pound a payment. The second is, that this being granted in this sort, other princes hereafter will look for the like; so we shall put an ill precedent upon ourselves and to our posterity; and in histories it is to be observed that of all nations the English care not to be subject, base, taxable, etc.

Bacon appreciated, ahead of many of his contemporaries, that the finances of the country were changing. Subsidies traditionally came largely from taxes on land, but an increasing proportion of the wealth of the country was associated with trade and industry, which was proportionally less taxed. The solution lay not in tripling the taxes on landowners, but in broadening the tax base. Nobody took much notice of this novel idea.

In the end, Parliament approved the subsidies, one to be paid in full in the first year, the second to be paid in full in the second year, and a third to be paid over the third and fourth years. The Queen had pretty much got what she wanted, and Bacon's opposition had never been to the subsidy but to the manner in which it was to be levied and the risk of establishing a precedent. Elizabeth, however, took offence. And, now in her sixties and somewhat crotchety, she certainly knew how to bear a grudge. The timing could not have been worse for

Bacon. The Master of the Rolls, Sir Gilbert Gerard, had recently died, and it was widely understood that he would be succeeded by the Attorney General, Sir Thomas Egerton – which would leave a vacancy for Attorney General, one of the plum jobs that Bacon craved. He marshalled all his resources, including both Essex and Burghley, to promote his claim to the place, but to no avail. The political manoeuvring dragged on until March 1594, with Bacon's hopes rising and falling over the months, before the Queen finally promoted Sir Edward Coke, the Solicitor General, to the post. This, of course, left a vacancy for Solicitor General, and Bacon and his supporters immediately took up his case as a candidate for that office. But after dragging the issue out until November 1595 (it is difficult not to believe that Elizabeth was deliberately toying with Bacon to punish him for his speech on the triple subsidy),[8] it was Sir Thomas Fleming who received the appointment. This was effectively the end of any prospects Bacon, now nearly 35 years old, had of preferment while Elizabeth remained on the throne. At least, though, that gave him more time to develop his ideas about how the proper application of knowledge might enable humankind to command nature in action. But it makes sense to at least sketch the outlines of the rest of his life in politics before we concentrate on his contribution to the birth of science.

The less Bacon pressed his claims for advancement, the more he returned to favour with the Queen, who made increasing use of his legal skills (sometimes even on a paid basis) and also treated him as one of her circle of advisers, although only on an informal basis. None of this did much to help Bacon's finances, though, and on one occasion, in September 1598 (just a month after his old mentor, Lord Burghley, died at the age of 74), he was actually arrested for debt. Although he managed to raise the funds to pay off this particular creditor, the incident drew uncomfortable attention to his impecunious state, and did nothing to ease his problems of obtaining further loans in the future.

8. Although the Queen may have been malicious, she was not stupid, and even during this period when Bacon was out of favour politically she continued to make use of his legal skills, and in July 1594 had appointed him as a member of her Learned Counsel, a post which conferred some prestige and the opportunity to carry out official commissions for the Queen, but usually on an unpaid basis.

But while Bacon's political star was at least modestly in the ascendant throughout the rest of Elizabeth's reign, Essex experienced a much more spectacular rise to glory, followed by an equally spectacular fall, in which Bacon again became highly visible, but this time on the side of the Crown against his former protector. Essex became something of a public hero in 1596, when he led an expedition that attacked Cadiz, destroying the Spanish fleet and reducing the risk of a Spanish invasion of England. Essex had wanted to follow up this success by an attack on the annual Spanish treasure fleet en route from South America, but was overruled by his more cautious (and more experienced) colleagues on the expedition, who were only too well aware of how much luck had played a part in their success. The Queen was less impressed, both by what she saw as Essex's self-promotion, and by the failure of the expedition to bring her any personal profit in the way of booty. While Essex had been away, the Queen had also appointed one of his rivals, Robert Cecil (by now, Sir Robert), as her Principal Secretary. Bacon went out of his way to advise Essex, still only 28 and with the prospect of a glittering future ahead of him, to bide his time and build cautiously on his military success, pointing out that a young military commander with popular support must be seen as a threat to an ageing queen with no direct heir, and advised him to avoid further military involvement and concentrate on securing his position at Court, perhaps seeking the post of Lord Privy Seal.

Far from following Bacon's advice, when a new Spanish fleet assembled at Ferrol, Essex eagerly sought, and obtained, the command of an English expedition to tackle it. This expedition, in 1597, suffered damage at the hands of a gale, and was left too weak to attack the main Spanish fleet. But Essex convinced himself and enough of his colleagues that the Spanish were not ready to leave Ferrol, and took his own fleet off on an unsuccessful attempt to find the latest Spanish treasure fleet (Sir Walter Ralegh was Essex's junior admiral on this expedition). The attempt failed, although they did at least bring back some booty from the Azores for the Queen's benefit, and while the English fleet was away the Spanish did leave Ferrol, with the intention of taking Falmouth, in the southwest of England, to use as a base. They would surely have succeeded if the fleet had not been overwhelmed by a huge storm, in which eighteen vessels were sunk.

As much by luck as judgement, Essex retained his reputation as a military leader, and he was appointed Earl Marshal of England (partly to balance the increasing power of the Cecils; Elizabeth always believed in keeping the different factions at Court busy watching each other to limit their opportunities to try to take control of her). It was in this capacity that Essex finally made his big mistake. In 1598, there was trouble in Ireland (there was always trouble in Ireland) where the Catholic population had religious reasons, as well as their natural objections to being a colony, to resent the English (it had only been in 1542 that Henry VIII had taken the title King of Ireland; before that, for four centuries English influence had been restricted to an area around Dublin, known as 'The Pale'). The latest wave of Irish resistance was led by the Earl of Tyrone, a talented politician and military leader educated in England, who had inflicted several defeats on English forces. Essex set off to subdue the Irish in the spring of 1599 (according to Bacon, very much against Bacon's advice), but after months of manoeuvring failed to bring the Irish to battle. Instead, he reached a verbal agreement with Tyrone, details of which were never committed to paper, and will never be known. By now, the Queen was suspicious of Essex's motives (by now, she was suspicious of everyone's motives), and ordered both him and his army to remain in Ireland until recalled. Against these orders, he returned to London, where he was placed under house arrest, while strange reports began to arrive from Ireland that Tyrone was boasting that he would soon 'have a good share in England'.[9] The inference was that Essex intended, using his army from Ireland and with Tyrone's support, to seize the English throne. In fact, it later emerged that Essex had been negotiating with the King of Scotland, James VI, as well, offering his support for the Scottish claim to the English throne (James Stuart was a descendant of Henry VII, Elizabeth's grandfather), in exchange, of course, for favours to himself.

Acting more generously than she might have, Elizabeth allowed Essex to return to more or less his normal life in London, while keeping him well away from the levers of power. Perhaps she was just giving him enough rope to hang himself; if so, she succeeded.

9. See Crowther.

He continued the secret negotiations with the Scots, and eventually developed a hare-brained plan to seize the Queen.[10] Early in 1601, it became obvious that something was afoot, and Essex was summoned to Court to explain himself. He then made a clumsy attempt to trigger a popular uprising, marching into the City of London with a couple of hundred armed supporters, mostly gentlemen. He found no support, went home, and was duly arrested and tried for treason.

The subsequent investigation and trial of Essex was carried out by a team of lawyers headed, of course, by the Attorney General, Sir Edward Coke. Bacon was a relatively junior member of this prosecuting team, but largely because of the bumbling way in which Coke handled the case, he found himself repeatedly standing up, in his clear and forthright manner, to keep the proceedings on track. He thus played a highly visible part in the conviction and subsequent execution (on 25 February 1601) of his former patron. Both at the time and later, Bacon was criticized for this, as if he owed loyalty to the man who had once looked after his interests at Court. But Essex had indeed committed treason, and Bacon's greater loyalty was to the Queen. He was, though, exasperated both by Essex's folly and by Coke's incompetence, infuriating reminders of how much better he could have done if given the same opportunities as either of them.

Bacon's personal fortunes began to change for the better at the beginning of the seventeenth century, although initially in unfortunate circumstances. In 1601 his brother Anthony, who had suffered a long illness, died, and Francis inherited the family home at Gorhambury. In the short term, this was a further financial encumbrance, because the house had not been well maintained, and in any case his mother still lived there. But it did offer the eventual prospect of a comfortable country retirement. Two years later, in 1603, Elizabeth died, having been born in 1533 and having reigned since 1558, for 44 years and 5 months. She was succeeded by King James VI of Scotland, who thereby also became James I of England. It would be under the Stuarts that Bacon would at last achieve all of his political ambitions, before

10. And the troubles continued in Ireland, with a Spanish invasion, supported by Tyrone, that was not defeated until the end of 1601. Tyrone did not submit until the end of 1602.

suffering his own fall from grace, although that could not have been predicted when James arrived in London with his entourage.

James was very much an established monarch, with his own way of doing things. Partly because of her Catholicism (and also because she was a rotten monarch) his mother, Mary Queen of Scots, had been forced to abdicate in his favour in 1567, when he was just a year old. The fact that James had managed to survive the subsequent turmoil in Scotland and emerge in charge indicates some political ability. Bacon, of course, was known to James chiefly as one of the prosecutors of Essex – one of whose crimes had been to collude with James. This was not an auspicious beginning, but his brother Anthony, a close associate of Essex, had been involved in the lengthy negotiations between England and Scotland about the Elizabethan succession, and this helped to ensure that Bacon was accepted at Court, where he retained his semi-official post as one of the 'Learned Counsel'. Initially, James had the good sense to maintain more or less the status quo at the Court in London, largely relying on Sir Robert Cecil for advice. Bacon seriously considered retiring from public life to concentrate on writing, to preserve for posterity the fruits of his philosophical musings. But as the consummate Parliamentarian he now was, he became deeply involved in drawing up plans for a political union of England and Scotland,[11] and through the friendship of Cecil and Cecil's influence over the King, he was knighted on 23 July 1603 – although to his chagrin this was as one of a job lot of some three hundred new knights created by the King as part of the celebration of his succession. But still, he was now Sir Francis, and the title would come in handy for a plan he had in mind.

In 1604, Bacon received his first official appointment, at the age of 43. He was made a King's Counsel (formalizing the status of Learned Counsel) at a stipend (then known as a pension) of £60 a year. In 1605, the year of the Gunpowder Plot, he published his *Advancement of Learning* (of which more later), dedicating it, with none too subtle flattery, to the King. And then, the Solicitor Generalship became vacant again. Bacon could not resist offering himself for the post, and

11. He also published, in 1603, *A Brief Discourse Touching the Happy Union of the Kingdoms of England and Scotland*. But these plans were premature; the Union itself, of course, did not take place for another century, until 1707.

while this was in the balance, the plan for which the knighthood was such an asset matured – he married. His chosen bride was Alice Barnham, the daughter of a wealthy alderman who had died in 1598 and left her £6,000 in land in Essex. Alice had been baptized on 4 June 1592, so cannot have been more than 12 years old when, in July 1603, Bacon had told Cecil, 'I have found out an alderman's daughter, an handsome maiden, to my liking';[12] she was probably about 14 when the marriage took place on 10 May 1606. Not much is known about Alice or the marriage, but it is clear (especially given his known sexual inclinations) that what was particularly to Bacon's liking about her was her dowry, which helped his finances enormously. But it is also clear that all was not plain sailing. Shortly before he died (when he was 65 and she was about 34), Bacon cut her out of his will, and eleven days after he died the young widow married John Underhill, who had been a member of Bacon's household (his post was 'Gentleman Usher'). It doesn't take a Sherlock Holmes to work out what had been going on under Bacon's roof.

But Bacon's rise continued after his marriage. He became Solicitor General (worth £1,000 a year) in 1607, and in 1608 the Clerkship of the Star Chamber at last reverted to him, bringing with it an income now increased to £2,000 a year. In 1613, Bacon was appointed Attorney General, in 1616 he became a member of the Privy Council, and in 1617 he became, like his father before him, Lord Keeper of the Great Seal. Just a year later, he was created Baron Verulam and given the title of Lord Chancellor; in 1621, shortly after his 60th birthday, he was created Viscount St Albans. And then, it all fell apart.

By 1620, Bacon – Lord Verulam, but we shall continue to refer to him as Bacon, for consistency – was one of the three most powerful men in the British Isles. James ruled as King of England, Scotland and Ireland, with Wales as a mere principality. He did so largely through his favourite, the Earl of Buckingham. Both relied almost entirely on the advice of the Lord Chancellor, who might reasonably be described as the brains of the outfit. But such a concentration of power inevitably provoked jealousies, and these powerful men had powerful enemies at Court and in the country at large. Of the three, Bacon was the

12. See Jardine and Stewart.

most vulnerable. Lord Chancellors could be replaced relatively easily, favourites (as the example of Essex showed) could be prised away from the monarch only with difficulty, and at that time the King himself was all but impregnable. The way to get at Bacon would be through Parliament; and in 1621 James called a Parliament (the first for seven years) chiefly in order to obtain funds to support his son-in-law, the Elector Palatine Frederick V of the Rhine, against the Spanish.[13] As ever, though, once a Parliament was called it tended to take the initiative in tackling other matters as well, and on this occasion there were genuine grievances about the way the triumvirate had been governing (particularly concerning the operation of monopolies; but the details need not concern us). The Lord Chancellor, of course, had been largely responsible for drafting the regulations which brought the unwelcome practices into force.

Bacon's enemies, among whom his old adversary Coke was prominent, sought for a chink in Bacon's armour through which the surge of ill-feeling against him might be directed. They found it in the workings of the Court of Chancery, over which he presided as Lord Chancellor. Witnesses came forward to the House of Commons to accuse Bacon of accepting bribes. But the issue is more complicated than this simple statement might appear to modern eyes. First, in Elizabethan (and earlier) times it was still the normal practice for anyone in high office to receive presents, perhaps in the form of cash or jewellery, from people they dealt with on official business. This was seen as some compensation for the fact that the offices involved were poorly paid, if at all. During Bacon's lifetime, this practice had begun to decline as society moved towards a system more like the modern one of paid officials; but he was now an old man, and old habits die hard. Secondly, the complaints were not so much that he had taken bribes, but that he had not been influenced by them! The accusers seemed to feel no shame at offering a bribe to sway a judicial decision their way, but they were angry that Bacon refused to be swayed. In which case, a lawyer might ask, could the presents really be regarded as bribes, rather than presents?

13. Frederick's grandson, George, would become the first Hanoverian King of Great Britain in 1714.

Because Bacon was now a peer, the Commons could not act directly against him, but sent their evidence to the House of Lords, as they put it (tongue firmly in collective cheek), 'without prejudice or opinion'. Bacon's first instinct was to defend himself against the charges, using in his defence the claim that he had accepted presents, but had not realized that they were intended as bribes (and so his judgement had not been affected by them), and also claiming that on many occasions the presents had been given to his servants, who had kept them for themselves without informing their master. Given the way Bacon's household was run (or not run), this was entirely plausible; some of his servants kept racehorses and retired to country estates, while Bacon himself, as we shall see, died technically bankrupt. Such a defence, though, drew attention to the irregular running of his household, which was hardly the approved way for Lord Chancellors to behave. For some reason, Bacon changed his mind, and decided to confess to the charges brought against him, converting this planned defence into a plea for mitigation. Nobody can be sure why he changed his mind, but an intriguing possibility was later suggested by his servant Thomas Bushell:

> There arose such complaints against his lordship, and the then favourite at court [Buckingham], that for some days put the King to this query, whether he should permit the favourite of his affection, or the oracle of his counsel to sink in his service? Whereupon his Lordship was sent for by the King, who after some discourse, gave him this positive advice: to submit himself to his House of peers, and that (upon his princely word) he would then restore him again, if they (in their honours) should not be sensible of his merits. Now though my Lordship saw his approaching ruin, and told his Majesty there was little hopes of mercy in a multitude when his enemies were to give fire, if he did not plead for himself, yet such was his obedience to him from whom he had his being, that he resolved, his Majesty's will should be his only law, and so took leave of him, with these words: 'Those that will strike at your Chancellor, (it's much to be feared) will strike at your crown'; and wished that as he was then the first, so he might be the last of sacrifices.[14]

14. Thomas Bushell, *Mr Bushell's Abridgement of the Lord Chancellor Bacon's Philosophical Theory in Mineral Prosecutions*, London, 1659.

In other words, the King sacrificed Bacon to appease his opponents and to save Buckingham. Although the account comes from a biased source, and may be unreliable, it would explain Bacon's change of mind, and would fit what we know of James's character (including the fact that when push came to shove his 'princely word' counted for nothing).

Bacon did not attend the proceedings in the House of Lords, where his confession and petition for mercy was presented on his behalf by James's son Charles, the Prince of Wales.[15] He was stripped of all his offices, of course, but allowed to keep his title as Viscount St Albans. He was fined £40,000, sentenced to be detained in the Tower 'during his Majesty's pleasure' (it soon pleased his Majesty to let Bacon return to his country retreat at Gorhambury, in June 1621), and was banned from holding any state office, sitting in Parliament, or coming within 12 miles of the Court. The enormous fine would have destroyed Bacon, but the King was persuaded not to levy it, even though it was still technically owed by Bacon. Even better, since the fine took precedence over all other debts, no other creditors could extract money from Bacon until the fine was paid – which it never was. Perhaps the princely word did count for something after all.

So Bacon was able to spend the last five years of his life in some comfort at Gorhambury (in January 1622, after intense lobbying, he even received a pension of £1,200 a year), where he devoted his time to writing and polishing the image which he wanted to show to posterity. His writings touched on history (including a history of the reign of Henry VII, James's ancestor), the law, philosophy, and what we would now call science. Only the science is relevant here, and the time has come to look at what Bacon really said, rather than what people in later generations thought he said.

Bacon's thoughts were developed over several decades, and although some fragments were published before 1620, there is much more unpublished material from the period before his fall from high office. We shall not attempt to follow the various steps and false starts

15. Charles only became heir to the throne when his elder brother, Henry, died of a fever (probably typhoid contracted from swimming in the Thames) in 1612 at the age of 18. In the light of subsequent events, one of the great 'what ifs' of history is, what if Henry hadn't taken that fatal dip? Would the relationship between Parliament and Henry IX have followed the same path as that between Parliament and Charles I?

he made while developing those ideas, but rather try to give you a feel for his final conception of how the investigation of the world ought to proceed. One thing is pre-eminent in all those writings, and was already clear in 1592, as the quote from *Mr Bacon in Praise of Learning* on page 62 highlights. Bacon thought that science[16] was pointless unless applied to the practical benefit of humankind. This is why, for example, he thought highly of Gilbert's work on magnetism, because it provided practical benefits for navigation and commerce, but was dismissive of his cosmological ideas, because they were of no practical benefit. It wasn't so much that Bacon thought these ideas were wrong (although he did scoff at the way Gilbert had taken one single idea, magnetism, and tried to explain everything in the Universe with it), but that there was no practical need of them. In the modern world, by analogy, we could imagine that Bacon would thoroughly approve of scientific investigation into the theory of flight, which has obvious practical benefits, but would scoff at the amount of effort devoted by scientists into investigating, say, the first split-second of the birth of the Universe. All this matters in terms of the way Bacon became a symbol for science among people who either did not know, or did not care, what he actually said. The Royal Society itself, although allegedly founded on Baconian principles, certainly never took upon itself any role in the practical application of science to the immediate direct benefit of humankind; if anything, it did the reverse, encouraging speculative investigation of the world by people interested in knowledge for its own sake, not for its practical utility.

It's perhaps worth mentioning in passing that Bacon's ideas about the dominance of humankind over nature were neither unique nor (even in 1592) entirely original. Giordano Bruno had been in England from 1583 to 1585, where he made quite an impact among the intellectual community in London, and published several pamphlets, in one of which he said:

> The gods have given man intelligence and hands, and have made him in their image, endowing him with a capacity superior to other animals. This capacity

16. We shall continue to use the modern term, to avoid repetition of the clumsy expression 'what we would now call science', although not without some slight misgivings. But it *was* science, in all but name.

consists not only in the power to work in accordance with nature and the usual course of things, but beyond that and outside her laws, to the end that by fashioning, or having the power to fashion, other natures, other courses, other orders by means of his intelligence, with that freedom without which his resemblance to the deity would not exist, he might in the end make himself god of the earth.[17]

It's easy to see why Bruno offended the Catholic establishment in Rome – and how he might have influenced the 24-year-old Francis Bacon. Much later, Bacon wrote:

Man is the helper and interpreter of Nature. He can only act and understand in so far as by working upon her or observing her he has come to perceive her order. Beyond this he has neither knowledge nor power.

The quote comes from Bacon's scientific masterwork, which was (or perhaps we should say, 'should have been') the *Instauratio Magna*, meaning 'Great Instauration' or 'Great Renewal'; it appeared in partial form in October 1620. The title alludes to the restoration of humankind, through the application of science, to the place occupied before the Fall described in the Bible – essentially the deification of humankind envisaged by Bruno. But the whole book was not published at the time – indeed, it never would be. What Bacon did publish was an outline of the great work he intended to write, and some parts of the planned masterwork, including a lengthy description of his method of research, under the title *Novum Organum*; as a result, the whole of the material published in 1620 is sometimes referred to by that title. The title itself alluded to Aristotle, who wrote a book called the *Organon*; the *Novum Organum* (*New Organon*) was intended to breathe new life into science by displacing the old Aristotelian approach.

The plan sketched out in the *Instauratio Magna* was in six parts. The first step in Bacon's scheme would be to compile a review of all existing knowledge; the second (the *Novum Organum*) described his scientific method; the third would be a natural history of the Universe; the fourth and fifth would provide results from the application of his method; and the sixth would be a philosophy derived from all the

17. See Waley.

preceding work (akin to what modern physicists call the 'theory of everything'). Before he died, Bacon himself compiled just three of the 126 natural histories he intended to form Part Three of the *Instauratio Magna*: the 'Winds', 'Life and Death', and 'Dense and Rare'. He always intended, and hoped, that other people would take up the task of completing his work. Indeed, one of the key points in his plan for science was the idea that it should be a cooperative venture, with discoveries shared freely among all scientists, not the kind of secret activity practised by the alchemists. His hopes for the completion of the *Instauratio Magna* were not fulfilled – at least, not in the way that he might have anticipated. There did grow up an encyclopedic tradition in France, aiming to classify nature as a prelude to understanding it, and this owed more than a little to Bacon, but is outside the scope of the present book.

As the emphasis on the natural histories shows, the key feature of Bacon's proposed scientific method was the collection of a vast number of facts, followed by an inductive approach to incorporating those facts into a theory – or theories – of the world. Broadly speaking, the inductive approach starts from generalities and moves towards specifics. The scientist collects data, as widely as possible, and then uses logical reasoning to generate a theory to explain the data. To take a very naive example, we might take a box containing a hundred apples, look at all the apples and see that they are red, and formulate a theory that all apples are red. The alternative approach, the deductive method, starts by identifying a problem, formulates a hypothesis which makes predictions about something that has not yet been observed, and collects data intended to test the accuracy of the predictions made by the hypothesis. The equivalent problem in the example we have already used would be to determine the colour of apples. Our hypothesis would be that all apples are red. We then test it by looking at apples, and are delighted to find that the first hundred we look at really are red.

The hypothesis becomes a theory if the results of this data collection (or experiment) match the predictions of the hypothesis. The deductive method starts from specifics and works outwards. But in both cases, the discovery of just one non-red apple pulls the rug from under the theory. Theories and hypotheses can always (potentially) be disproved in science; but they can never be proved beyond all doubt; there is always the possibility that one more experiment may turn up the equivalent of a non-red apple. There is room for both approaches in modern science; even this crude characterization of the two methods shows that the way science actually developed was far from being exclusively through the approach Bacon envisaged.

The thing that looks particularly weird to modern eyes is that Bacon was trying to develop a method for doing science in a step-by-step approach that would lead even the humblest practitioner to the correct conclusions, with no need for geniuses or flashes of insight – no room, in fact, for the imagination. He made an analogy with the difficulty of drawing a straight line freehand, which required a steady hand and a great deal of practice, and drawing a straight line with the aid of a ruler, which anyone could do. His method was to be to science what a ruler was to the drawing of straight lines, a mechanical aid to discovery, or an inductive engine. 'My way of discovering sciences[18] goes far to level men's wits,' he wrote, 'and leaves but little to individual excellence; because it performs everything by the surest rules and demonstrations.' And he illustrated the steps involved in this process with an example of the investigation of the nature of heat.

First, he collects all the examples he can think of of the phenomenon being investigated (in this case, heat) in a 'Table of Existence and Presence'. This includes 'the rays of the sun particularly in summer, and at noon', but also 'severe and intense cold' because that produces 'a sensation of burning'. Next, he constructs a 'Table of Deviation', listing things that in some way resemble those in the first table, but do not show the phenomenon being investigated – so 'the rays of the moon, stars and comets' are included here, because although in some ways like the rays of the Sun, they 'are not found to be warm to the touch'. A third table records variations in the degree of the phenomenon in the same thing at different times, and in comparison with other objects. At last, armed with all this information, you are supposed to be ready to start the process of induction, using a method of Exclusion to determine what aspects of the phenomenon are not essential to its nature – for example, the fact that the Sun's rays are hot excludes the

18. Remember that in Bacon's day 'science' meant 'knowledge'.

possibility that heat is a purely terrestrial phenomenon. Suddenly, however, in a completely baffling step Bacon tells us to jump from this step-by-step approach to the formation of a hypothesis, which he calls the 'first vintage', in anticipation that it will be modified in the light of further observations and experiments:

> Since, however, truth emerges more readily from error than confusion, we consider it useful to leave the understanding at liberty to exert itself and attempt the interpretation of nature in the affirmative, after having constructed and weighed the three tables of preparation, such as we have laid them down, both from instances there collected, and others occurring elsewhere. Which attempt we are wont to call the liberty of the understanding or the commencement of the interpretation of the first vintage.

In other words, the time has come for a leap of the imagination, a flash of insight, or a stroke of genius! We might have understood better how Bacon made such an inspired leap, and the role it played in his method of induction, if he had lived to write Part Four of his great work, where he promised to demonstrate the system more fully; but all he ever wrote of this part of the *Instauratio Magna* was the Introduction. Wherever it came from, Bacon's own insight, or hypothesis, based on his tables concerning heat, is that

> Heat is an expansive motion restrained, and striving to exert itself in the smaller particles. The expansion is modified by its tendency to rise, through expanding towards its exterior; and the effort is modified by its not being sluggish, but active and somewhat violent.

Ignore the fact that this is remarkably like the modern kinetic theory of heat. What matters is that there is no way in which such a hypothesis could have been derived simply by the method of observation, negation, and exclusion. It is a guess – based on those tables, to be sure, but not an inevitable consequence of them. Once you have the guess, or hypothesis, it can be tested by experiment and observation in the usual way (and if necessary refined into a second vintage, and so on). But at the heart of Bacon's own method, unacknowledged, is the fact that progress in science depends on the application of the imagination. It is *not* a purely mechanical process that can be worked out by anyone by following a recipe book. But Bacon does rightly emphasize the importance of testing hypotheses and falsifying them as a route towards ultimate truth, which really is a key step in the scientific method. It is no good just pointing to the occasions when a hypothesis is supported by the observations; you must also consider the failures, as he highlights with a little parable:

It was well answered by him who was shown in a temple the votive tablets suspended by such as had escaped the peril of shipwreck, and was pressed as to whether he should then recognise the power of gods, by an enquiry: But where are the portraits of those who have perished in spite of those vows.

The point about the Baconian method is that Bacon intended it to have widespread – perhaps universal – application to problem solving. Trivial examples, like our box of red apples, make it seem almost like a tautology or common sense, like the story of the man who 'had been speaking prose all his life, and never knew it'. This left Bacon open to ridicule in some quarters, and a notable example comes from Thomas Macaulay's *Essay on Bacon*, where he describes the hypothetical application of the inductive method by a man who eats minced pies and finds they disagree with him. It runs as follows:

'I ate minced pies on Monday and Wednesday, and I was kept awake all night.' This is the *Table of Presence*. 'I did not eat any on Tuesday and Friday, and I was quite well.' This is the *Table of Absence in Proximity*. 'I ate very sparingly of them on Sunday, and was very slightly indisposed in the evening. But on Christmas day I almost dined on them, and was so ill that I was in great danger.' This is the *Table of Degrees*. 'It cannot have been the brandy which I took with them. For I have drunk brandy daily for years without being the worse for it.' This is the *Table of Exclusion*. Our invalid then proceeds to what is termed by Bacon as the *First Vintage*, and pronounces that minced pies do not agree with him.

Macaulay intends to poke fun at Bacon by demonstrating that his method of induction is something we do all the time, and that Bacon did not invent it. In fact, he provides us with a nice, clear example of how the method works, and misses the point that what Bacon was trying to do was to formulate this almost instinctive method of human reasoning into a system which could be applied to much more complicated situations. An appropriate analogy might be with the way

Aristotle did not discover, or invent, logic, but did analyse this method of reasoning and set out the rules it obeyed for others to follow. It is Macaulay, not Bacon, who emerges with egg on his face.

But this doesn't mean that Bacon really had hit on 'the' way to do science. Quite apart from the need for the leap of the imagination, the big – almost insurmountable – difficulty facing anyone trying to apply the strict Baconian method in any non-trivial situation is knowing when to stop listing things and making tables, and when (and how!) to make the imaginative leap to a hypothesis of the first vintage, always assuming you have a spark of the right kind of imagination needed to make such a leap. In the trivial example used above, do we stop after a hundred apples, or a thousand, or a million? Or do we go on forever? I can think of only one important figure in the history of science who really did work along Baconian lines (and he died in 1601, long before Bacon spelled his ideas out), and another who had Baconian leanings. The Danish astronomer Tycho Brahe carried out endless (in the sense that they continued until he died) observations of the motion of heavenly bodies, in particular Mars, without ever taking the step of guessing what those observations were telling us. It was left to Johannes Kepler, in the next generation, to use these data to confirm his speculative idea that the orbit of Mars traced an ellipse around the Sun – a profound break with the ancient belief that because circles are perfect, and the heavens are perfect, the orbits of heavenly bodies must be circular. The semi-Baconian scientist was, of course, Charles Darwin, who spent a vast amount of time collecting and comparing specimens of different species in a way that closely matched the ideas behind Bacon's system, but who then made the spectacular leap to the theory of natural selection to explain his observations. In most other cases, the speculations stemming from the intuition or imagination of great scientists were based on relatively little evidence, and grew in importance only because they were then tested experimentally and by observation, and passed those tests (the intuitive hypotheses that failed these tests have been consigned to the dustbin of history and forgotten). But there is a role, as Bacon clearly realized, for people who do not rank as geniuses, or have the spark of scientific imagination, to do what might be called the spadework of science, the collecting and enumerating and the preparation of the tables.

All this explains why Bacon's contemporary William Harvey, who happened to be Bacon's own doctor and also discovered the circulation of the blood (see Chapter Three), described Bacon as writing philosophy (that is, philosophy of science) 'like a Lord Chancellor'.[19] He meant that Bacon wrote like a legislator, laying down a rigid procedure to be followed, like the letter of the law. Another person who wrote about science like a Lord Chancellor was the philosopher Thomas Kuhn, who (like Bacon) had no experience of scientific research himself but promoted the idea that science proceeds through a series of revolutionary changes. As I have discussed elsewhere,[20] and as even the example of Gilbert's work on magnetism shows, this is a complete nonsense, and science really proceeds incrementally, building step by step on what has gone before. But there are many non-scientists, and even some scientists, who believe that Kuhn's ideas about scientific revolutions and 'paradigm shifts' must be true, because they are written in a book. In a classic example of the way science really works, as we shall see, Harvey himself intuited that blood must circulate on the basis of relatively slender evidence, then carried out the observations and experiments which confirmed his hypothesis (or rather, did not refute it). But even he did not go the whole revolutionary hog and leap to the conclusion that the heart was merely the biological equivalent of a mechanical pump.

The only other major part of the *Instauratio Magna* that Bacon completed was Part One, published in 1623 under the title *De Dignitate et Augmentis Scientiarum* (*On the Dignity and Advancement of Learning*). This was essentially an expanded and updated version of his 1605 publication *Advancement of Learning*, in which he spells out the idea of scientific (technological) progress as a steady advance, with the lot of humankind being improved as time passes on an endless climb towards the kind of terrestrial deification that Bruno had envisaged. This may seem obvious today. But Aristotle, in his *Metaphysics*, said explicitly that the process of discovery had already been completed, that all the necessities and refinements of life had

19. See Aubrey. King James, who thought of himself as an intellectual, and sponsored the classic English translation of the Bible, described Bacon's philosophy as 'like the peace of God, which passes all understanding'; see Farrington, *Francis Bacon*.
20. *Science: A History*.

already been invented; and Plato, although not spelling this out so explicitly, had the same view. As late as the sixteenth century, with Aristotelian ideas holding sway, this was still a widespread view, in spite of the technological changes that had already begun to take place. Bacon saw this blind adherence to outmoded ideas as a major obstacle to the kind of progress he envisaged, because it implied that the best that could be hoped for was to maintain the status quo:

> There is a powerful cause why the sciences have made but little progress. It is not possible to run a course rightly when the goal itself has been wrongly set. Now the true and lawful goal of the sciences is simply this, that human life be endowed with new discoveries and powers.

But it should be emphasized that Bacon fully appreciated that in the wrong hands scientific discoveries and technological inventions could be put to evil uses. In a rather laboured analogy based on the story of Daedalus and the labyrinth he constructed to conceal the Minotaur, Bacon concludes:

> Certainly human life is much indebted to [the mechanical arts], for very many things which concern both the furniture of religion and the ornament of state and the culture of life in general, are drawn from their store. And yet out of the same fountain come instruments of lust, and also instruments of death. For (not to speak of the arts of procurers) the most exquisite poisons, as well as guns, and such like engines of destruction, are the fruits of mechanical invention.

Clearly, though, to Bacon this is no reason to give up the pursuit of knowledge, any more than the fact that knives can be used to kill is a reason to stop making them.

In order to achieve the great objectives he envisaged for science, Bacon appreciated the practice of science would have to become a professional, state-supported activity, with suitable colleges, libraries, experimental facilities, and well-funded expeditions to seek out new things. Above all, he saw science as a collective effort. To a large extent, although he was not a scientist himself, he created the image of what a scientist ought to be, and thereby helped to pave the way for developments later in the seventeenth century. The sort of thing he had in mind was spelled out in his unfinished book, *New Atlantis*, published shortly after his death in an edition edited by Rawley, which had a great influence on the founders of the Royal Society.

This fable describes a mysterious island in the uncharted regions of the South Pacific, which is found by a ship that has got lost. The island, called Bensalem, turns out to be inhabited by an advanced civilization descended from the survivors of the inundation of Atlantis (hence the title), with a utopian society ordered around the work of a scientific organization called the Foundation:[21]

> The End of our Foundation is the knowledge of Causes, and secret motions of things; and the enlarging of the bounds of Human Empire, to the effecting of all things possible.

To this end, the Foundation has established large orchards and gardens, kept for the cultivation of new kinds of plants, pastures where new varieties of animals were bred, powerful and effective medicines, 'divers mechanical arts, which you have not', and much more besides. They can sort light into its component colours, and present them 'not in rainbows, as it is in gems and prisms, but of themselves single'. Bacon clearly appreciated the limitations of the human senses in the investigation of the secrets of nature (hardly surprising, in an age when the telescope and microscope had begun to be used by scientists), and wrote elsewhere that 'by far the greatest hindrance and aberration of the human understanding proceeds from the dulness, incompetency and deceptions of the senses', something the Foundation overcame with artificial aids. The members of the Foundation, who apply this knowledge for the benefit of the people of Bensalem, call themselves Fellows; it was in homage to Bacon that the members of the Royal Society called themselves Fellows, with other learned societies later following their lead. He provided them, and us, with a vision of what the world could be like if science were indeed applied to the service of humankind.

But before we look at how the turbulent changes that took place in English society in the decades following Bacon's death paved the way

21. *New Atlantis* has been called an early example of science fiction; this is a moot point, but it is true that Isaac Asimov took the name 'Foundation' from Bacon for his eponymous series of stories.

for the Royal Society to (as they would have seen it) pick up the baton from him, I should take the opportunity to lay to rest the popular, but apocryphal, tale of how Bacon himself met his death – if only because I have previously been guilty of spreading the popular myth myself. This legend, which we get from Bacon's former secretary Thomas Hobbes, via the always entertaining but not always reliable Aubrey, has it that Bacon was 'taking the air' in a coach at the beginning of April 1626, in the company of one of the King's physicians, a Dr Witherborne. He suddenly took it into his head to seize the opportunity provided by the snow lying about to carry out an experiment in preserving a dead chicken. Leaping from the coach, he purchased a hen, had it gutted, and used his own hands to stuff the body with snow. As a consequence of these exertions in the cold, he caught a chill, took refuge in the Earl of Arundel's nearby house at Highgate, and died because he was put into the best guest room, where the bed had not been used for a year and was damp.

It's a delightful tale, but scarcely matches what is known about Bacon's character. Jardine and Stewart have constructed a convincing alternative scenario, starting from the observation that Bacon was already ill by the end of 1625, and sufficiently unsure of his future that on 19 December that year he revised his will to cut his wife out of the inheritance. He was certainly too sick to be 'taking the air' in the snows of early April 1626, and must have had a strong motive to be out and about. A clue to that motive, and just how the illness got worse, was provided by Bacon himself in a letter he dictated from his sickbed to the Earl of Arundel (who was not in residence in Highgate) explaining why he had imposed himself upon the household:

I was likely to have had the fortune of Caius Plinius the elder, who lost his life by trying an experiment about the burning of the mountain Vesuvius. For I also was desirous to try out an experiment or two, touching the conservation and induration of bodies. As for the experiment itself, it succeeded excellently well; but in the journey (between London and Highgate) I was taken with such a fit of casting [vomiting], as I knew not whether it were the stone, or some surfeit, or cold, or indeed a touch of them all three. But when I came to your Lordship's house, I was not able to go back, and therefore was forced to take up my lodging here, where your housekeeper is very careful and diligent about me; which I assure myself your Lordship will not only pardon towards him, but think the better of him for it.

... I know how unfit it is for me to write to your Lordship with any other hand than mine own; but in troth my fingers are so disjointed with this fit of sickness, that I cannot steadily hold a pen.

As well as indicating that he was unaware that his own illness really was terminal, Bacon's joking reference to Pliny the elder (who was suffocated by fumes on the slopes of Vesuvius) indicates that the experiments he referred to, aimed at prolonging life, had involved inhaling vapours, and had been carried out in London. An interest in the prolongation of life is a thread running through much of Bacon's writings, and something he became increasingly interested in as his own body aged. Although never as ill as his brother Anthony, who died young, Bacon had never enjoyed the rudest health, and like many of his contemporaries was something of an addict of quack remedies, opiates and the chemical saltpetre (or nitre) which was usually, in those days, found as a mixture of potassium nitrate, sodium nitrate, and various impurities. In his *History of Life and Death* we find the claim that 'the human spirits can be cooled and condensed by the spirit of nitre, and made more crude and less eager. Nitre composes and restrains the spirits and tends to longevity.' The inference drawn by Jardine and Stewart, from this and other evidence, is that Bacon had been inhaling both nitre and opiates, in an attempt to 'indurate', or strengthen, his own failing body, and to prolong his own life – this would also explain the presence of Dr Witherborne. After an initial drug-induced high ('it succeeded excellently well') he had been hit by the inevitable consequences as he came down from that high on the road back home. And one of the side effects of an overdose of such inhalations, which Bacon had mentioned in his *Life and Death*, is that they 'mortify' the extremities of the body, such as the fingers, so that it would be difficult to hold a pen.

The most likely pattern of events is that in his weakened state Bacon had taken far too much 'physik' for his body to cope with. Whether or not a damp bed then played a part, he died on the morning of Easter Sunday, 9 April 1626, leaving debts of nearly £20,000. By then, Charles I was already on the throne (he had succeeded his father in 1625), and the next step forward in the development of science in England had already been taken, and was about to be published, by William Harvey, who was a physician not only to Bacon, but to both James I and Charles I, and whose life takes us right through the period of the Civil War, and up to the first stirrings of the movement that gave birth to the Royal Society.

# 3

# Making the Transition

William Harvey was a transitional figure in several ways. Chronologically, he linked the time of Gilbert, Galileo, and Bacon with the period of the Parliamentary Interregnum, and the first stirrings of the movement that would become the Royal Society; scientifically, although his own discoveries were based on experiment and observation, he saw himself as building on the Aristotelian tradition, not as a pioneer of a completely new way of investigating the world; and even geographically his own career exemplifies the way in which the centre of scientific gravity in Europe was shifting, in the first half of the seventeenth century, from Italy to England.

We know very little about Harvey's personal and private life. The only piece of information that has come down to us about his wife, for example (apart from her name and the fact that the couple had no children), is that she kept a tame parrot. Fortunately, though, enough is known about his professional career and place in society to paint a picture of his life outside the home. Harvey was born on 1 April 1578 at Folkestone, in Kent. His father, Thomas Harvey, was a yeoman farmer who owned land on the downs above Folkestone (overlooking the modern site of the entrance to the Channel Tunnel), and the family seem to have lived in the area and worked the land for several generations before him. In 1576, when he was 25 years old, Thomas married Juliana Jenkin, the daughter of William Jenkin and Mary Halke, but she died the following year, almost certainly in giving birth to her only child, a daughter. Within a year, Thomas Harvey had remarried, to a Joan Halke who was probably the cousin of his first wife (the Halkes were a prolific and prominent local family). William was the eldest of seven sons and two daughters produced by the

*William Harvey*

couple; all his brothers went on to successful careers – five in business in London and the other at Court – another indication of the rise of the middle class in England in Elizabethan times.

William Harvey was sent to study at the King's School, in Canterbury, in the summer of 1588, the year of the Spanish Armada, when he was 10. The school vies with the Cathedral School in York to be regarded as the oldest school in England; it had originally been a cathedral school for boys intending to join the monastery, but under the Reformation of Henry VIII it had become a secular school, under its new name, in 1541. There is no record of Harvey as a boarder at the school, and most probably he lived with the family of Thomas Jenkin (a prosperous merchant who was a brother of Thomas Harvey's first wife) while he was in Canterbury. Geoffrey Keynes[1] has pointed out that one Robert Halke, probably another relative of young William, lived at the time in the house next door to the playwright Christopher Marlowe in Canterbury. Robert Halke was an apothecary, and it is possible that this family connection triggered William's interest in medicine. But he would certainly not have learned anything about science or medicine at school, where the traditional grounding in the Classics, Latin and Greek, formed the bulk of the syllabus.

Whatever the reason, though, it is clear that Harvey had developed an interest in medicine by the time he left school, in 1593, to go up to Cambridge University at the age of 16. He chose to go to Gonville and Caius College, which was then the leading place to study medicine in England.[2] This was thanks to the influence of Dr John Caius, who had studied in Padua alongside Andreas Vesalius (of whom more shortly) and on his return to England had essentially taken over the old Gonville Hall, a minor Cambridge institution which he had attended as a student, providing endowments and re-founding it as Gonville and Caius College. He was Master of the new college from 1549 until 1573, shortly before he died; he would no doubt be delighted to learn that to this day it is usually referred to simply as Caius. One of his achievements as Master was to obtain from the

1. See Sources and Further Reading.

2. Cambridge University as a whole was something of an academic backwater in Elizabethan times so the medical tuition available at Caius College was an isolated example of a centre of excellence.

Queen a charter entitling the college to be provided with the bodies of two executed criminals each year for the purpose of dissection.

This slightly gruesome note provides an opportunity to point out that a great deal of Harvey's own later work (and, of course, the work of his contemporaries) depended not just on dissection of fresh human corpses, but on experiments carried out on living animals – vivisection. This would be completely unacceptable today, and the details of some of the experiments are not for the squeamish (don't worry, we won't go into them here). But attitudes towards animals were very different in those days – the 'lesser' animals were seen as having been put on Earth for the benefit of humankind – and even the attitude towards human life (as shown by the kind of crimes for which the penalty was death) was more casual. Attitudes were simply different then; everyone in Elizabethan England (especially the son of a farmer) would have been well aware where the meat that they ate came from, and would not be cushioned from reality like the modern shopper pushing a trolley round the supermarket. Reporting the results of those experiments is not intended to condone the practice of vivisection or to suggest that they would be acceptable today, any more than a report of the family lives of the rulers of Ancient Egypt would be intended to condone or encourage incest.

There may have been another reason for Harvey to choose Caius as his college. In 1571, the then Archbishop of Canterbury, Matthew Parker, had founded a medical scholarship worth just over £3 per annum (very nearly enough to live on in those days) for the benefit of boys born in Kent and educated at the King's School in Canterbury. This was the first medical scholarship in England (undoubtedly extracted from Parker by his friend John Caius), and William Harvey became its beneficiary (the fourth student to do so) in December 1593, soon after joining the college. All the evidence is that he was a serious student who had little time for frivolity. Harvey took his BA in 1597, when he was 19, but stayed in residence in Cambridge, receiving his stipend as Matthew Parker Scholar and (presumably) studying until December 1599. During this period he suffered a severe bout of illness, which has been tentatively identified as malaria, rife at the time both in Kent and in the marshy fen country around Cambridge.

Early in 1600, Harvey travelled to the Italian city of Padua to continue his studies at what was then the best medical school in Europe.[3] The transition must have been easier than it would be today in one regard – as one of the few things he would have learned at school was a fluent grasp of Latin, and Latin was the teaching language in all the great universities in Europe, he would not have had to learn a new language before attending lectures (although it is likely that he would have had to get used to the different way the Italians pronounced the common tongue). Padua was the best university in Italy (and therefore possibly in the world) at this time, and benefited from being close to the Republic of Venice; it was officially regarded as the Venetian university, and Venice was the most liberal and open-minded of the Italian city states at the beginning of the seventeenth century. Giordano Bruno, who was burned in Rome the year Harvey arrived in Padua, had been a professor at the university. Whereas the English universities, Oxford and Cambridge, were run by the Fellows of the colleges (senior members with at least an MA to their names), at Padua the student body elected representatives to the council which supervised the teaching, and they also elected their own teachers. Each national group at the university constituted a college (or 'Nation'), and Harvey, although a newcomer to Padua, was elected *Consiliarius* for the English Nation in August 1600; he retained this status in each of the next two years. He was there just at the right time to benefit from the first new ideas in medicine to have been developed in over a thousand years.

In the second half of the sixteenth century, medical knowledge – and in particular the understanding of human anatomy – had begun to make its first real progress since the time of Ancient Rome. The last word in Roman understanding of how the human body is put together had been written by Claudius Galenus (actually a Greek physician, usually known as Galen), who had been active in the second half of the second century AD. Galen pulled together the ideas of his predecessors (notably Hippocrates and Aristotle), and was a skilled

3. Galileo was in the middle of his happy spell as a professor in Padua while Harvey was there, but this was just before he became famous through his telescopic observations. There is no evidence that the two ever met. 1600 was also, of course, the year Gilbert published *De Magnete*, but Harvey rose through the ranks of physicians only after Gilbert's death; again, there is no evidence that the two actually met.

anatomist who carried out his own dissections. Unfortunately, because the religious views of the period precluded human dissection, many of his conclusions about human anatomy were based on dissection of other animals, including apes, and were therefore incorrect. One passage from his writings indicates the difficulty he worked under, and the frustration he must have felt:

It is possible to see something of human bones. I have done so often on the breaking open of a tomb or grave. Thus once a river, inundating a recently made grave, broke it up and swept away the body of the dead man. The flesh had putrefied, but the bones still held together in their proper relations.

Galen's ideas about blood saw it as part of a system of four 'humours' which determined the health of the body. These were blood, phlegm, yellow bile, and black bile. The idea in ancient times was that if the four humours were in balance, the body was healthy; but if they were out of balance, in one way or another, the body fell sick. It was this kind of reasoning that led to the practice of bleeding a patient – if the explanation for an illness was that there was an imbalance caused by an excess of blood compared with the other humours, then the logical thing to do was to drain some of that blood away.

Galen thought that there were two distinct kinds of blood in the human body. This was not a completely mad idea; blood from arteries is a bright scarlet colour (we now know, because it is rich in oxygen) while blood from veins is dark purple (because it lacks oxygen). Arteries themselves are quite different from veins – arteries have thicker, more elastic walls than veins, and they pulse, while veins do not. At first sight, the two kinds of blood look like different fluids, carried in different kinds of piping, and that is what the Ancients thought them to be. One kind was thought to be manufactured in the liver (as a result of the digestion of food) and carried throughout the body by the veins, to provide nutrition. On this picture, the nutritive blood was constantly being manufactured, and constantly being used up, in a one-way flow outward from the liver. The second kind of blood carried the spark of life ('vital spirit') around the body. Air from the lungs, it was thought, became mixed with blood, and this 'vivified' blood was passed to the left side of the heart and from there distributed to the rest of the body through the arteries. Again, this was seen as a steady process, with the vivified blood being constantly used up and replaced; in this case the fresh supply of blood was thought to come via the heart, where blood from the liver entered the right-hand side of the heart, passed through hypothetical tiny holes in the central wall of the heart, and travelled from there to the lungs to be vivified. Apart from the fact that there is no hole (or holes) connecting the right and left chambers of the heart, if you replace the term 'vital spirit' with 'oxygen', and appreciate that the blood circulates, this isn't all that far from the truth. The modern view is that veins carry blood towards the heart, arteries carry blood away from the heart, and it keeps going round the body – but nobody before Harvey appreciated that the blood does circulate.

Obviously, though, even on Galen's picture the blood did move through the veins and arteries. Galen thought that the heart sucked blood towards itself when it was expanding, and that its contraction was simply a relaxation before the next expansion; in the same sort of way, he thought that the pulse attracted blood from the heart to where the vital spirit was needed. This wasn't really the kind of 'attraction' that you would produce by sucking liquid up through a straw but something more mystical; the important point is that on this picture the pulsebeat and the heartbeat would be out of step with one another.

No progress was made in medicine for the next 1,300 years. If anything, even what little had been known in Roman times was largely forgotten in Europe, although Galen's writings, like many other Classical texts, were preserved first in Byzantium and then in the Arab world. Within the overall Renaissance that began in the fifteenth century, medicine began to make its mark when the works of Galen were published in a new Latin translation (from the original Greek) in 1525, and disseminated widely thanks to the printing press. At first, like so much Classical knowledge in the early Renaissance, this was treated like some kind of holy writ that could never be improved; but soon one man, Andreas Vesalius, was questioning the accuracy of Galen's teaching and taking anatomy, in particular, further (although Galenist ideas held sway in some quarters well into the seventeenth century).

Vesalius came from Brussels, where he had been born in 1514, and studied at Louvain and in Paris before arriving in Padua (where one of his contemporaries, as we have seen, was John Caius) to complete his medical education in 1537. Unlike Galen, Vesalius was able first to see human dissections carried out, and then to carry out his own dissections; but the way anatomy was taught still left much to be desired. The teaching (such as it was) was based on the assumption that Galen was right, and that the purpose of the dissection was to point out the truths he had laid down. The professor in charge of the dissection would read the relevant passages from a text (probably Galen), while a surgeon (in those days, a very lowly member of the pecking order) carried out the actual dissection, and a third member of the team, called an ostensor, would use a pointer to indicate the various organs and so on being referred to by the professor. The idea was simply to demonstrate what was already known, and had been known since Galen's time.

Vesalius, thanks to the education he had already received in Louvain and, in particular, Paris, almost immediately received his degree of Doctor of Medicine, and became a professor himself. But he did things differently. He carried out the dissections himself, while explaining to the students what it was he was uncovering, instead of reading from a book about what Galen said they ought to be uncovering. This example began to spread to other universities, much to the benefit of anatomy. And it soon became obvious to Vesalius where Galen had gone wrong in extrapolating from his ape dissections to describe human anatomy. In 1543 (the same year that Copernicus published *De Revolutionibus*) he published a gorgeous, beautifully illustrated book on human anatomy, *De Humani Corporis Fabrica* (usually referred to as the *Fabrica*), which stressed the importance of carrying out dissections with your own hands and looking at the evidence with your own eyes, rather than relying on the wisdom of the Ancients. Then, at the age of 29, Vesalius gave up his position in Padua and spent the rest of his life as a Court physician, first to Charles V, the Holy Roman Emperor, and then to Philip II of Spain (the same Philip who sent the Armada against England). But his influence at Padua remained strong. One of his students, Gabriele Fallopio (also known as Gabriel Fallopius, the discoverer of the 'Fallopian tubes'), became Professor of Anatomy in Pisa in 1548, but came back to Padua as

Professor of Anatomy in 1551. Although Fallopio died in 1562 at the age of 39, he was succeeded in the post by his own former pupil, Girolamo Fabrizio (who, following the fashion for Latinizing names, became known as Hieronymus Fabricius ab Aquapendente), a surgeon brought up in the Vesalian, not the Galenic, tradition. It was Fabricius who taught Harvey, and who made a key discovery which helped lead Harvey to discover the circulation of the blood.

Fabricius had been born in 1537 – as his name suggests, in the town of Aquapendente, in Umbria. He was a committed teacher, who had a special lecture theatre constructed to his own design to make it easier for students to see what was going on at dissections, and held the Chair of Anatomy at Padua until 1613; he died in 1619. When Harvey arrived in Padua, Fabricius was in his 63rd year, but still an active and enthusiastic teacher, and in his masterwork, *Exercitatio Anatomica de Motu Cordis et Sanguinis in Animalibus*,[4] Harvey refers to him as 'the celebrated Girolamo Fabrizi d'Acquapendente a most skilful anatomist and venerable old man'. Some of Fabricius' own work concerned embryology and the development of the foetus in hen's eggs, which, as we shall see, may also have influenced Harvey; but his most important contribution was a clear and accurate description of the valves in veins which, we now know, prevent the blood in them from moving away from the heart. These valves – actually flaps rather like trap doors – had been seen by earlier anatomists,[5] but it was Fabricius who shone a spotlight on them and triggered Harvey's awareness of their importance. According to his own account, Fabricius first noticed the valves during a dissection he carried out in 1574, and he was certainly describing them in public demonstrations and his lectures at the university from 1579 onwards; he did not write down what he had discovered, though, until 1603, in a pamphlet published under the title *De Venarum Ostiolis*. The title itself shows that Fabricius did not appreciate just what he had found. Remember

4. Usually referred to as *De Motu*; the title translates as *Anatomical Exercises on the Motion of the Heart and Blood in Animals*, but is sometimes shortened in English translations to *Movement of the Heart and Blood*. The reference to Fabricius as if he were still alive is one of the clues that *De Motu* was actually largely completed before the end of 1619.

5. Including Paolo Sarpi, who features in Galileo's story.

that at that time it was still thought that the flow of blood in veins was *away* from the heart. The valves, or flaps, completely block the passage of blood in this direction, but allow essentially unimpeded flow towards the heart. *Ostiolum* means 'little door', and from the context it is clear that Fabricius thought of the flaps as resting slightly open, like a door that is ajar, to slow down (but crucially, not to halt) the flow of blood away from the heart. There was a seemingly obvious reason why this might be important in the veins of the legs, for example, where gravity would be pulling the blood downwards, and the little doors would stop it flowing too fast and swelling up the veins in the feet. It was Harvey, in *De Motu*, who introduced the Latin term *valvula* to describe the little flaps, and said that 'the discoverer of the valves did not understand their real function, and others went no further'.

But just how soon did Harvey understand their function? Much later, Robert Boyle (one of the key figures in the story of the Royal Society) described a conversation in which he

> Asked our famous *Harvey*, in the only Discourse I had with him, (which was but a while before he dyed) What were the things that had induc'd him to think of a *Circulation of the Blood*? He answer'd me, that when he took notice that the Valves in the Veins of so many several parts of the Body, were so plac'd that they gave free passage to the Blood Towards the Heart, but oppos'd the passage of the Venal Blood the Contrary way: He was invited to imagine, that so Provident a Cause as Nature had not so Plac'd so many Valves without Design: and no Design seem'd more probable, than That, since the Blood could not well, because of the interposing Valves, be Sent by the Veins to the Limbs; it should be Sent through the Arteries, and Return through the Veins, whose Valves did not oppose its course that way.[6]

The point is that if the blood is flowing from the extremities to the heart, it must have got to the extremities in the first place through the only other route possible, the arteries. This passage from Boyle is sometimes taken to imply that Harvey experienced a moment of epiphany as a student, watching Fabricius demonstrate the existence

6. Quoted by Keynes, from Boyle's *A Disquisition about the Final Causes of Natural Things*, published in 1688.

of the valves and seeing the truth in a blinding flash. But there is nothing in this passage to tell us *when* Harvey 'took notice' of the importance of what he had seen demonstrated as a student, and all the other evidence suggests that it happened much later, long after he left Padua. It makes sense to complete our sketch of his life story, before coming back to the importance of his work, and his book, in aiding the transition from Aristotelianism to science.

Harvey received his Doctor of Medicine degree in Padua on 25 April 1602. By the summer, he was back in England, where he took up residence in London and began the process of applying for membership of the College of Physicians (later to become the Royal College), without which he could not practise; he never bothered to obtain an MD from Cambridge, but in 1642 he was incorporated as MD at Oxford when in residence there as the King's Physician. The College took its time about admitting new members, and although Harvey made his first application on 4 May 1603, when he faced an oral examination, the records of the College note that although 'his replies to all questions were entirely satisfactory . . . Nevertheless he was put off until another time, with our tacit permission to practise.' Just under a year later, on 1 April 1604, Harvey was examined again and approved; this was followed by a third examination on 11 May, and finally on 7 August he was elected to membership of the College; he paid his dues and took the required oath on 5 October, becoming free to practise medicine officially within the jurisdiction of the College, with the status of Licentiate. Meanwhile, no doubt on the strength of his successful candidature, having obtained a marriage licence on 24 November the same year, he married Elizabeth Browne, the 23-year-old daughter of a London physician, Lancelot Browne; but, as we have said, apart from the bare fact of the wedding and the passing reference to Elizabeth's parrot in Harvey's writings,[7] we know nothing else about Harvey's wife.

His father-in-law, though, was certainly a useful connection. Lancelot Browne was a senior member of the College of Physicians, where he had served as a Censor (that is, Examiner) and was on the Council in 1605. Like William Gilbert, he was one of the physicians

7. In *De Generatione Animalium*, published in 1651.

at the Court of Queen Elizabeth, and after her death in 1603 held the same post under King James. As early as July 1605, Browne put Harvey's name forward when the post of Physician at the Tower of London became vacant; but the application was not successful. Browne died in December 1605 (a few weeks after the death of Harvey's mother), and Harvey was left to make his own way as a physician, without patronage. He practised medicine privately in London for the next few years, and in 1607 was elevated to the status of Fellow of the College; but by 1609 he had established another connection at Court, through his younger brother John, who had obtained a post described as 'footman' to the King. It isn't clear now just what duties this entailed, although it is clear that the post was more prestigious than that of a modern-day footman. Perhaps because of this connection, in February 1609 Harvey was able to make his application for the post of Physician at the Hospital of St Bartholomew's 'in reversion', accompanied by a letter of recommendation from the King. This was rather like the way Bacon had been granted the reversion to the Star Chamber post – if the application succeeded (which it did) Harvey would be sure of the post at St Bartholomew's when the current incumbent died, or retired, and the hospital could be sure there would be a smooth and speedy transition. Meanwhile, he could learn the ropes at the hospital by standing in for the present Physician, a Ralph Wilkinson, when he was away on other business. As it happens, Wilkinson died in the autumn of 1609; we don't have the details, but clearly he had been ill, and Harvey's application for the post was made in the expectation that it would soon become vacant. The post provided him with an income of £25 per year, and 40 shillings a year for his livery; he later received further payment in lieu of taking up residence in a house provided by the hospital. Perhaps more significant than the income, there were many fine houses owned by gentlemen high in Stuart society in the vicinity of the hospital, providing Harvey with more contacts. One of these neighbours, Sir Ralph Winwood, is known to have entertained on a lavish scale, and in April 1617 Bacon, recently appointed as Lord Keeper, was the guest of honour. Harvey must have shared in the social life of his neighbours, although whether he was present at that particular dinner is a matter of conjecture.

He was certainly starting to treat a better class of patient by then, including, in 1612, the Lord Treasurer, Robert Cecil. He was also moving up the hierarchy at the College of Physicians, and becoming more involved in its administration, initially as a Censor in 1613. Even so, it was also towards the end of the second decade of the seventeenth century that he carried out most of the research which led him to discover the circulation of the blood. In 1615, Harvey was appointed as Lumleian Lecturer by the College.[8] The post, founded by John, the sixth Lord Lumley, in 1582, required its holder to give a series of anatomy lectures, and the notes for those lectures are among the few items of Harvey's that have survived to the present day. Although these contain no reference to the circulation of the blood (except for a page that was obviously added some years later), they provide insights into Harvey's character and his way of reconciling the things he saw with his own eyes with the establishment view that Galen was right in all things. As he wrote in those notes, he aimed 'to learn and teach anatomy, not from books, but from dissections, not from the positions of the philosophers but from the fabric of nature'.

At one point,[9] describing the way skin thickens and becomes calloused by work, he uses as examples the flanks of horses toughened by constant spurring, and (tongue clearly in his cheek) that of saints' knees, thickened by constant kneeling in prayer. Referring to critics of Galen who say that they have found only four lobes in the lungs of humans rather than the five he described (five lobes are found in apes), he says, 'I suspend my censure . . . for perhaps in Galen's time it was common in man whereas now it is rare.' The suggestion that the structure of the human body had changed so dramatically in a couple of thousand years, if intended seriously, is breathtaking; the obvious interpretation is that Harvey was giving his audience the choice of deciding for themselves whether such a transformation was more likely than the possibility that Galen had in fact been dissecting apes, not people. Elsewhere in the notes, he refers to the ability of the human colon to pass objects such as lead bullets and 'my own gold

8. The first of his Lumleian lectures was delivered in 1616, the week before William Shakespeare died.
9. See Whitteridge.

ring', suggesting that in the great tradition of medical experimenters he had used himself as an experimental subject.

But although in 1616 Harvey had not yet realized that the blood circulated through the body as a whole, his lecture notes show that he had begun to develop a clear understanding about how the heart worked and that, drawing on the work of his predecessor Realdus Columbus, he was one of the first people to appreciate properly and accept fully the idea of what is known as the 'lesser circulation' in which blood travels from the heart to the lungs and back again. Blood which enters the right-hand side (ventricle) of the heart from the body via the veins is pumped to the lungs along the pulmonary artery (an artery because the blood in it is flowing away from the heart; but in this case only that arterial blood is oxygen-depleted). Then the blood absorbs oxygen before flowing back into the left-hand side of the heart along the pulmonary vein (a vein because the flow is towards the heart, even though in this particular vein the blood is oxygen-rich). From here, we now know, the oxygen-rich blood is pumped through the arteries to the body. That is the modern picture, including oxygen, which was only discovered in the 1770s. Following Galen, in the sixteenth century it was thought that blood got to the left-hand side of the heart from the right through tiny holes in the wall separating the two chambers of the heart, where it mixed with air brought from the lungs along the pulmonary vein, and became imbued with 'vital force' before travelling into the body through the arteries. The tiny holes in the heart wall had to be there for this idea to work, but it became increasingly difficult to believe in them as time passed. In 1543, Vesalius wrote in his *De Humani Corporis Fabrica*:

> The septum of the ventricles, therefore, is, as I have said, made out of the thickest substance of the heart and on both sides is plentifully supplied with small pits which occasion its presenting an uneven surface towards the ventricles. Of these pits not one (at least as far as is perceptible to the senses) penetrates from the right ventricle to the left, so that we are greatly forced to wonder at the skill of the Artificer of all things by which the blood sweats through passages that are invisible to sight from the right ventricle to the left.

By the time he wrote the second edition of 1555, his doubts had grown:

Howsoever conspicuous these pits may be, not one of them, in so far as is perceptible to the senses, penetrates through the ventricular septum from the right to the left ventricle. Indeed, I have never come upon even the most obscure passages by which the septum of the heart is traversed, albeit that these passages are recounted in detail by professors of anatomy seeing that they are utterly convinced that the blood is received into the left ventricle from the right. And so it is (how and why I will advise you more plainly elsewhere), that I am not a little in two minds about the office of the heart in this respect.

But even Vesalius never accepted the idea of the lesser circulation.

The idea seems to have occurred independently to at least two people, and possibly three. As far back as the thirteenth century, the Arab scholar Ibn al-Nafis came up with a version of the idea, but his work was unknown to the physicians of the sixteenth and early seventeenth centuries, and had no influence on Harvey. The first European to publish the idea of the lesser circulation was the Spaniard Michael Servetus, in 1553. Servetus was a theologian and physician, born in 1511, who presented the idea in passing in a theological treatise, *Christianismi Restitutio*; it was not based on dissection and observation, but on theological and mystic arguments. Servetus' theology (he was an anti-Trinitarian, and did not believe that Jesus Christ was God incarnate) got him into trouble with the authorities, and he was burned at the stake in Geneva later in 1553. As all but a handful of the copies of his book were also burned, he played no further part in the story of the circulation of the blood – indeed, the first known reference to his work wasn't published until 1694.

It isn't clear, though, that Servetus really did hit on the idea entirely independently, since in the early 1550s Realdus Columbus, in Italy, was already discussing the idea, and it is just possible that word of this may have got to Servetus. Columbus himself, who lived from 1510 to 1559 and was Vesalius' immediate successor in Padua, although he left in 1547, did not write about the lesser circulation until the publication of his book *De Re Anatomica* (dated 1559, but not actually in circulation before 1560); however, his pupil Juan Valverde (another Spaniard) had referred to it and attributed the discovery to Columbus in a book, *De la composición del cuerpo humano*, published in 1556. The details may never be untangled, but it is clear that Harvey got the idea solely from Columbus, and knew nothing of either Ibn al-Nafis or Michael Servetus.

Columbus had reached his conclusions as a result of three straightforward observations. There were no holes in the wall dividing the two halves of the heart; there was only blood in the pulmonary vein, not air; and there were very efficient valves (more complicated and efficient than the relatively simple flaps found in the veins) in the heart which only allowed a one-way flow of blood outward through the pulmonary artery and inward through the pulmonary vein. The other key insight contributed by Columbus, and taken up by Harvey, was that the heart is working when it squeezes itself down, pushing blood out along the arteries, and that this is what produces the pulse. Harvey's lecture notes do not specifically refer to the lesser circulation, but his references to Columbus show that he had read his book and must have been aware of the idea. The notes also show that in 1616 he was thinking about the role of the valves:

> The arteries have a thicker coat, and especially in adults whose heart beat is stronger, because the artery sustains the impulse . . . The pulmonary artery is thicker for it sustains the pulse of the right ventricle in adults and of the artery in the foetus. Hence also, neither the vena cava nor the pulmonary vein is of such a structure because they do not pulsate, but rather they may be said to suffer an attraction and this because the valves set in a contrary direction break off the pulse both in the heart and in the other veins . . . the veins have very many valves opposed to the heart, while the arteries have none except in the exit from the heart in contrary fashion.

It seems likely to us that it was while working on his Lumleian lectures that Harvey first 'took notice' of the important clue provided by the valves in the veins, and that his ideas about the circulation of the blood were developed over the next few years, when he was in his late thirties and early forties. This ties in with Harvey's statement in *De Motu*, published in 1628 (when Harvey was 50 years old), that the demonstrations on which his ideas were based had been confirmed 'for nine years and more'. Columbus would have been delighted, not just to see his ideas taken up and used to advance human knowledge but by the way Harvey based his own discoveries on careful experiment and observation. Lashing out in frustration at the traditionalists who regarded it as impossible to improve on Classical knowledge, in his own book Columbus had written:

Whosoever is willing to consider these reasons with an open mind, will, I know well enough, agree, and will allow a place to be given to the truth, even though Galen, the great philosopher and chief of all physicians, if we except Hippocrates, does in fact seem to have been ignorant of this use of the lungs. Let it be, he is a great philosopher and a still greater physician, and yet it is not surprising that these things and many others have been concealed from man. Yet truly, there is a race of men stupid, and ignorant, who have neither the wish nor the ability to find anything new. And therefore, whatever a physician with a great name writes, they immediately subscribe to it, nor will they depart from their beliefs not one jot. But you, honest Reader, a pupil of learned men and most zealous for the truth, I beg of you try the experiment in living animals . . . I admonish you, I exhort you, try the experiment I say, and find out whether what I have said agrees with the thing itself.[10]

'Try the experiment . . . and find out whether what I have said agrees with the thing itself.' There is the scientific method encapsulated, in a book published in 1559. But the fact that the first person who really did try the experiment, not only confirming the discoveries made by Columbus but taking things further, was William Harvey, some sixty years later, shows how slowly the scientific 'revolution' got started.

But while Harvey was quietly making his own contribution to the scientific revolution, his status as a prominent member of the College of Physicians was still growing – between 1607 and 1628 he held every office in the College except that of President. By 1618 (the exact details of the appointment have been lost), he had become one of the physicians to King James, in addition to his duties at St Bartholomew's, his private practice, and his administrative work with the College. It's a reminder of how small the circles of Stuart society were that at this time the King's chaplain (appointed in 1615) was the renowned churchman and poet John Donne; the two can hardly have failed to meet, and Donne is known to have had an interest in medicine (his stepfather was a physician). But it is astonishing to find him preaching

10. Quoted by Whitteridge, *William Harvey*.

a sermon at Whitehall on 8 April 1621 in which he says (before going on to contrast the finite limits of the body with the infinite possibilities of spiritual things):

We know the receipt, the capacity of the ventricle, the stomach of man, how much it can hold; and wee know the receipt of all the receptacles of blood, how much the body can have; so wee doe of all the other conduits and cisterns of the body.

The astonishing thing is that one of Harvey's key arguments in *De Motu*, as we shall see, concerns the amount of blood in the body, and the capacity of the heart. Donne must have got hold of the idea either from Harvey personally, or by attending some of his lectures; from this and other evidence it seems clear that as well as his public lectures Harvey also gave private demonstrations for his interested friends and colleagues, where he went further in discussing the circulation of the blood than the surviving notes for his official lectures suggest.

As if he wasn't busy enough already, in 1620 Harvey became involved in an investigation of the ancient monument of Stonehenge. The King had got interested in the stones while visiting the Earl of Pembroke at his residence near Salisbury, and sent his architect, Inigo Jones, to survey the site. Jones concluded that the site was the ruin of a Roman temple, but didn't publish anything about his investigations. After Jones died (in 1652) his pupil John Webb published Jones's ideas, in 1655. This provoked a response from Walter Charleton (another royal physician) attacking the idea, which in turn provoked Webb to publish a response to Charleton. Our interest in the story is that in his second book, published in 1665, Webb mentions Harvey as one of the people who persuaded him to publish the original account of Jones's work, and goes on to describe 'Heads of Bulls, or Oxen, of Harts, and other such Beasts', found at the site and attributed to Roman sacrifices. 'The abundance of them which were digged up by Dr Harvey formerly mentioned, Gilbert North Esquire, Brother to the Right Honourable Lord North, Mr Jones, and divers other persons.'[11]

James I, not yet 60 years old, died at the end of March 1625. His

11. See Keynes.

death was attributed by Harvey, who attended him during his final illness, to 'ague', an all-purpose term to refer to feverish maladies, and possibly referring to malaria. Rumours that the King was poisoned by his adviser, the Duke of Buckingham, seem to have been unfounded, but when (in July 1626) the new King, Charles I, gave Harvey a gift of £100 'for his pains and attendance about the person of his Majesty's late dear father', it was accompanied by a 'general pardon' (similar to the ones issued these days by outgoing American presidents) for anything he might have done during the previous reign. This was probably only meant to ensure that Harvey could not be blamed for anything he had done on behalf of James I; it may have been intended also to imply the confidence the new King had that Harvey, at least, had behaved correctly during the last illness of King James.

At the end of 1627, Harvey became one of the Elect, the highest rank in the College of Physicians short of the Presidency. A year later, *De Motu* was published. And all the while Harvey's friendship with the new King, who had been 24 on his accession, was increasing. Charles had initially been very much under the influence of Buckingham (giving some weight to the rumours about Buckingham's role in the death of his father), who was a self-seeking politician whose influence led England into disastrous military adventures against Spain and France (the latter even though Buckingham himself had arranged the marriage between Charles and the French princess Henrietta Maria, a Roman Catholic). In 1627, Buckingham was in command of an army sent to aid the French Huguenots in La Rochelle, but was defeated in a four-month campaign. When Parliament urged the King to dismiss Buckingham, Charles refused. The Duke then (in August 1628) began to assemble another expeditionary force in Portsmouth, where he was assassinated by John Felton, a navy lieutenant who had served under Buckingham and knew what another French adventure would involve. In London, the populace rejoiced at the news. Charles, freed from Buckingham's influence, turned to his wife for emotional support, and his family became the centre of his life.

In 1629, the year after the publication of *De Motu* (to which we shall return later), Harvey was serving the College as both its Treasurer and as a Censor, but because of his strengthening association with the King, these would turn out to be the last posts he held there. In

December that year he resigned both positions because he had been asked by the King to accompany James Stewart, the young (nearly 18) Duke of Lennox and a relative of Charles I, on a trip to the continent. It was also necessary to appoint a deputy who could stand in for him at St Bartholomew's. The Duke visited France and Spain over the next couple of years, but cut short plans for a journey to Italy because of an outbreak of plague in the regions he would have had to travel through. Harvey certainly was not with the Duke throughout his continental excursion, since he is recorded at various times attending the College of Physicians; but he may have been away from London in the summer of 1630 or during the following winter. Whatever services he provided for the Duke, in December 1630 the King was pleased to raise his status from Physician Extraordinary (which, in spite of the way modern readers might read such a title, meant he was just one of the physicians called on by the Court) to Royal Physician in Ordinary (which meant that he was one of the King's two personal physicians, junior to a Dr Bethune). This was a big change in Harvey's life. He could no longer hold office at the College (although he remained an Elect), and the centre of his life shifted to the Court, where he now received an annuity, nominally of £50 per year although this was paid erratically in arrears and often in larger lumps. It also meant having a permanent deputy, Dr Andrews, at St Bartholomew's, although Harvey retained his official status there and continued to receive his annual fee.

In 1633, Harvey accompanied the King on the first visit Charles had made to his native Scotland since coming to London with his father in 1603. A year later, he was involved in one of the periodic outbursts of hysteria about witchcraft that marred those times. The details of the accusations need not concern us, but when Harvey was called in to examine four condemned witches, his cool appraisal of their condition was largely responsible for them being pardoned. Their accuser later confessed that he had made up the stories about witchcraft. The significance of the story to us is that it demonstrates Harvey's logical, scientific approach at work outside the dissecting room; his duties inside the dissecting room included (in 1635) carrying out a post-mortem examination of Thomas Parr, a poor farmer known as Old Parr, who had died at the alleged age of 152 years and

9 months. A more thankless task which fell to Harvey the following year was to treat the King's headstrong nephew, the 21-year-old Prince Charles Louis, who largely ignored Harvey's advice, as he ignored all good advice during his life.

Charles Louis was the eldest son of Charles I's sister Elizabeth, who had married Frederick V, the Elector Palatine of the Rhine;[12] his brother Prince Rupert, born in 1619 (the year before the *Mayflower* sailed for the New World), would distinguish himself fighting for the King in the English Civil War. In the turmoil of European politics at the time (roughly in the middle of the Thirty Years War), Charles Louis was shunted out of the succession to the Palatinate when in 1635 Emperor Ferdinand of Germany (a Catholic) struck a deal giving the title to the Duke of Bavaria, who was also a Catholic and just happened to have married a daughter of the Emperor. This wasn't an issue that justified England going to war, but Charles I had to do something for his nephew, and decided to send a large embassy, headed by Thomas Howard, the Earl of Arundel, to plead his case with the Emperor. Arundel chose Harvey as the physician to accompany the expedition. The party left London on 7 April 1636 for what turned out to be a lengthy excursion taking in Regensburg, Vienna, and Prague. Harvey met many prominent physicians and anatomists on his travels, and in July he was given leave to make a private journey into Italy, travelling as far as Naples, while the Earl's party waited for the opportunity to press Charles Louis's case with the Emperor. He rejoined the Earl in November, when Arundel was ready to return home, his mission (as expected by everyone except Charles Louis) having failed. They were back in England before the end of the year.

Little is known about Harvey's life over the next couple of years, but it seems likely that he accompanied the King on at least some of the expeditions to the north to suppress the Covenanters, Scottish Presbyterians who objected to the imposition of bishops on them and opposed the idea of the divine right of kings. In July 1639, Dr Bethune

12. The Electors were a group of German princes and archbishops who elected the Emperor of Germany (the Holy Roman Emperor). It was through this branch of the Stuart line that the Hanoverians gained their claim to the throne of England.

died, and Harvey was promoted to become the senior Physician in Ordinary, bringing him an income of £400 a year and entitling him to move (which he promptly did) into lodgings in the royal palace of Whitehall. In 1640 the King was back in the north of England to counter a Scottish invasion which reached as far south as Newcastle and Durham; the resulting peace treaty, signed in October, favoured the Scots. But this was far from being the end of the King's troubles. In November, what became known as the Long Parliament met in London for the first time. This was the Parliament with which Charles would soon be in conflict.

The Parliament had been called for the usual reason – Charles had been bankrupted by what became known as the Bishops Wars, and by a renewed outburst of trouble in Ireland, calling for yet more military intervention, and needed money. He got it, but only by giving concessions to Parliament in turn. He also needed to re-establish what passed for friendly relations with the Scots, and in the summer of 1641 he made another visit to Scotland, again with Harvey in attendance. By giving out new titles and allowing an increased measure of self-government, he seemed to have succeeded. But while he was away Parliament passed the Grand Remonstrance, which condemned 'in detail and at length the King's policy in Church and State, at home and abroad, throughout his reign'. On 10 January 1642, the King and his family left London for Hampton Court, where they were soon joined by Harvey. After Queen Henrietta Maria left to seek help from the French, the King's entourage travelled north to York to begin raising an army; the English Civil War formally began when Charles raised his standard at Nottingham on 22 August that year.

The Civil War had one immediate, and personally disastrous, effect on Harvey. His quarters in Whitehall were ransacked, in what seems to have been a deliberate act of vandalism, and his papers destroyed, including notes he had gathered with the intention of writing more books. When he did publish his other great work, *De Generatione Animalium*, in 1651, he lamented in it:

My misfortune: namely, that while I did attend upon our late Sovereign in these late distractions, and more than Civil Wars; (and that not by the Parliaments bare permission, but command) some rapacious hand or other not onely spoiled me of all my Goods; but also (which I most lament) have bereft me of my Notes, which cost me many years industry. By which means, many observations (especially those concerning the *Generation of Insects*) are lost and imbezzled, to the prejudice (I may boldly say it) of the Commonwealth of Learning.

Apart from this, though, Harvey had what would now be described as a good war, especially considering that he was on the losing side. At the Battle of Edgehill, in October 1642, Harvey was still with the King, and so were the two royal princes, the Prince of Wales (later Charles II), then aged 12, and the Duke of York (later James II), then aged 9. Harvey was given charge of the two boys, and during the long morning while the opposing armies prepared for battle he sat with them under a hedge, where he calmly read a book until the cannonballs started flying.[13] Well, he was 64 by then, and had seen enough of life not to get too excited at the prospect of the forthcoming battle.

With the battle inconclusive, and winter coming on, Charles decided to set up his Court in Oxford (where 16-year-old John Aubrey was in residence at Trinity College). It was then, on 7 December 1642, that Harvey was incorporated MD at the university. He almost immediately received bad news about his twin brothers, Matthew and Michael, who died within a few weeks of each other, Matthew on 21 December and Michael on 22 January 1643. While the King continued his struggle against the forces of Parliament (who were soon to be joined, decisively, by the Scots), Harvey was based in Oxford for the next three and a half years, becoming Warden of Merton College and continuing his research into embryology, but also travelling on various medical duties for the King. While at Merton, Harvey became the friend and to some extent patron of Charles Scarburgh, a medical student who had left Cambridge for Oxford because of his Royalist sympathies; the connection is noteworthy because in 1647, after Harvey had been ousted from Oxford, Scarburgh in his turn took an interest in the career of a brilliant 15-year-old student, Christopher Wren. In a sense, Harvey was Wren's academic grandfather. But

13. See Aubrey. Aubrey had the account of this adventure directly from Harvey, and there is no reason to doubt it.

his relatively peaceful interlude in Oxford came to an end when Parliamentarian forces took the city in June 1646, and Royalists became personae non gratae.

By this time, the King was being held in Newcastle as a prisoner of the Scots, and asked for Harvey, his physician, to be allowed to visit him there. Permission was granted, and Harvey was allowed to travel north at the end of the year, presumably staying with the King until he was moved to Holmby, in Northamptonshire, early in 1647. From then on, the King was forbidden to choose his own doctors or chaplains. When negotiations for a deal between Parliament and the King broke down, and the King engineered a Scottish invasion of England *against* Parliament by promising the Scots that he would establish Presbyterianism in England, the Civil War entered its second bloody phase, culminating in the defeat of the Scots at Preston in August 1648, and the execution of Charles I on 30 January 1649. Harvey, stripped of all his posts by the Parliamentarians, and now in his 70th year, had returned to London in January 1647, to live in retirement with his surviving brothers, Daniel, who had a house in Croydon, and Eliab, who had homes at Roehampton and at Chigwell. He continued to practise medicine for a few select patients, including a Mr Montague who was the cousin of John Aubrey and through whom he made Aubrey's acquaintance. Another close friend, remembered in his will, was the philosopher Thomas Hobbes, ten years younger than Harvey, and a friend (or at least acquaintance) who was also one of his patients was Robert Boyle, one of the founders of the scientific movement in England. So Harvey knew both Fabricius (born in 1537) and Boyle (who died in 1691).

In 1654, Harvey declined the Presidency of the College of Physicians on the grounds of age and infirmity – it was probably never expected that he would accept, but the Elect wanted to do him the honour of offering the post. As you might expect from his long association with the sick and his experience of human dissection, Harvey was unsentimental about the prospect of dying, and he never seems to have been overtly religious. The circumstantial evidence lends credence to a story preserved in the diary of Viscount Perceval, who became the first Earl of Egmont, and lived from 1683 to 1748; the entry is dated 20 June 1740, and reports a story told to him by Dr Edward Wilmot, a Fellow of the Royal College of Physicians and one-time friend of Charles Scarburgh:

[He] told me an anecdote of the famous Dr Harvey, the discoverer of the circulation of the blood, namely that he voluntarily killed himself with laudanum, being one of those whom, if he were now living, we should call a free thinker, and who believed it lawful to put an end to his life when tired of it. The first attempt he made to do it was unsuccessful, as Dr Scarborrow, his intimate friend, related it, who agreed in opinion with the other that suicide is lawful. One day, Harvey being in great pain (he was then about 72 yrs old) sent for Scarborrow, and acquainting him with his intention to die by laudanum that night, desired he would come next morning to take care of his papers and affairs. Scarborrow, who had long before promised him that friendly office when occasion called on him, did accordingly come next morning, but was surprised to find Harvey alive and well; it seems the laudanum he had taken, instead of killing him, had brought away a considerable number of stones, which effect caused a suspension of his design to destroy himself for some years.[14]

When Harvey did die, on 3 June 1657 at the age of 79, rumours that he had committed suicide were sufficiently widespread that Aubrey felt he had to scotch them:

Had he laboured under great paines, he had been readie enough to have donne it . . . but the manner of his dyeing was really, and *bona fide*, thus, viz. the morning of his death about 10 a'clock, he went to speake, and found he had the dead palsey in his tongue; then he saw what was to become of him, he knew there was then no hopes of his recovery, so presently sends for his young nephewes to come-up to him, to whom he gives one his watch ('twas a minute watch with which he made his experiments); to another, another remembrance, etc.; made sign to . . . his apothecary (in Black-Fryers) to lett him blood in the tongue, which did little or no good; and so he ended his days.

Aubrey attributed the source of the rumour to Charles Scarburgh, who had said that Harvey had

Towards his latter end, a preparation of opium and I know not what, which he kept in his study to take, if occasion should serve, to putt him out of his

14. See Keynes.

paine, and which Sir Charles Scarborough promised to give him; this I believe to be true; but doe not at all beleeve that he really did give it him . . . The Palsey did give him an easie passe-port.

The great achievement of Harvey's 'retirement' was the publication of his *De Generatione Animalium*, in 1651. This impressive study of embryology and animal reproduction hints at what must have been lost when the Parliamentarians ransacked Whitehall in 1642; but whatever else he might have written, *De Motu* would always have been regarded as Harvey's masterwork. Before we look in detail at how Harvey applied the scientific method to discover the circulation of the blood, however, there is one curious and irresistible story that Harvey recounts as an aside from his main theme in *De Generatione Animalium*.[15] The events he describes took place some time around 1640, when he examined the Irish nobleman Hugh Montgomery, who was in his late teens at the time. Montgomery had suffered a serious accident in his childhood, and made a bizarre recovery:

A Noble young gentleman, Son and Heire to the honorable the Vice-Count of Montgomery in Ireland, when he was a childe, had a strange mishapp by an unexpected fall, causing a Fracture in the Ribs on the left side: the Bruise was brought to a Suppuration, whereby a great quantity of putrified matter was voided out, and this putrefaction gushed out for a long while together out of the wide wound . . . This person of Honour, about the eighteenth, or nineteenth year of his Age, having been a Traveller in Italy and France, arrived at last in London: having all this time a very wide gap open in his Breast, so that you might see and touch his Lungs (as it was believed). Which, when it came to the late King Charles his ear, being related as a miracle, He presently sent me to the Young Gentleman, to inform Him, how the matter stood. Well, what happened? When I came neer him, and saw him a sprightly youth, with a good complexion, and habit of body, I supposed, some body or other had framed an untruth. But having saluted him, as the manner is, and declared unto him the Cause of my Visit, by the Kings Command, he discovered all to me, and opened the void part of his left side, taking off that small plate, which he wore to defend it against any blow or outward injury. Where I

15. The quote here is taken from the English language version, *Anatomical Exercitations Concerning the Generation of Living Creatures*, published in 1653.

presently beheld a vast hole in his breast, into which I could easily put my three Fore-fingers and my Thumb: and at the first entrance I perceived a certain fleshy part sticking out, which was driven in and out by a reciprocal motion, whereupon I gently handled it in my hand. Being now amazed at the novelty of the thing, I searched it again and again, and having diligently enough enquired into all, it was evident, that that old and vast Ulcer (for want of the help of a skilled Physitian) was miraculously healed, and skinned over with a membrane on the Inside, and guarded with flesh all about the brimmes or margent of it. But that fleshy substance (which at the first sight I conceived to be proud flesh, and every body else took to be a lobe of the Lungs) by its pulse, and the differences of rythme thereof, or the time which it kept, (and laying one hand upon his wrest, and the other upon his heart) and also by comparing and considering his Respirations, I concluded it to be no part of the Lungs, but the Cone or Substance of the Heart; which an excrescent fungous Substance (as is usual in foul Ulcers) had fenced outwardly like a Sconce. The Young Gentleman's Man did by daily warm injections deliver that fleshy accretion from the filth & pollutions which grew about it, and so clapt on the Plate; which was no sooner done, but his Master was well, and ready for any journey or excercise, living a pleasant and secure life.

So Harvey actually saw, and handled (albeit through the membrane that had healed over the inside of the wound), a living human heart. He was also able to confirm from direct observations that the pulsebeat was in step with the contraction of the heart which forced blood into the arteries and around the body. As for Montgomery, he continued to lead an active life, fighting on behalf of the King in the Irish troubles, and against Parliament in the Civil War. Although banished to Holland after the Parliamentary victory, he was returned to Ireland as Master of Ordnance by Charles II, but didn't live long to enjoy this status as he died suddenly in 1663. Alas, history does not record if his death was linked to the hole in his chest.

Apart from its curiosity value, the story shows Harvey the scientist (or proto-scientist) at work. For what must have been at least ten years a succession of people, surely including other doctors, had looked at the pulsating flesh revealed by the cavity in Montgomery's chest and guessed that they could see the lungs at work. Harvey, however, observed carefully and measured the rhythm of these pulsations, comparing them with the pulsebeat in Montgomery's wrist and the beating of his heart heard, or felt, in the usual way through the chest wall. It was by the same kind of careful observation and measurement that he found evidence for the circulation of the blood.

When Harvey published *De Motu*, he must have had at least tacit approval from the College of Physicians, and it may be that one reason for the delay of nine years or so in publication was because of the time it took for him to persuade his peers of the soundness of his work. In a preface to his book, Harvey uses the conventional device of an open letter to the then President of the College, John Argent, to explain his hesitancy and justify his ultimate decision to publish:

> Since this Book alone does affirm the blood to pass forth and return through unwonted tracts, contrary to the received way through so many ages of years insisted upon and evidenced by innumerable, and those the most famous and learned men, I was greatly afraid to suffer this little Book, otherwise perfect some years ago, either to come abroad, or go beyond the Sea, lest it might seem an action too full of arrogancy, if I had not first propounded it to you, confirm'd it by ocular testimony, answer'd your doubts and objections, and gotten the President's verdict in my favour; yet I was perswaded if I could maintain what I proposed in the presence of you and our College, having been famous by so many and so great men, I needed so much the lesse to be afraid of others, and that only comfort, which for the love of truth you did grant me, might likewise be hoped for from all who were Philosophers of the same nature.

Even the way the book was published shows Harvey as a transitional figure, since he chose to send the manuscript to the publisher William Fitzer, in Frankfurt. The main reason for this was to ensure a wide readership for the book; in the 1620s, a book published in London would be unlikely to receive as much attention as one published on the continent. Before the end of the century, the situation would be reversed, with scholars on the continent sending their papers to the Royal Society in London for publication, to ensure maximum publicity for their work. A secondary consideration in Harvey's case may have been the fact that although the English publishers would expect an author to subsidize the cost of printing his work, Fitzer actually offered not only free author's copies of the book, but payment as well.

The downside of this decision was that it was a long way to Frankfurt, and communications between London and Germany were hampered by the ongoing wars. So Harvey (who had atrocious handwriting) had no chance to correct proofs, and the first edition of his masterwork is littered with misprints. The little volume (just 68 pages of text) was also published on cheap paper (Fitzer had to cut corners somewhere if he was going to pay his authors) which crumbled away sooner rather than later; only 55 copies survived into the second half of the twentieth century. But by then, of course, it had been corrected and reprinted in many editions.

There are two classic pieces of evidence that Harvey uses in *De Motu* to make his case, and which highlight the scientific approach. Since we have already mentioned the importance of the valves in the veins, let's start with them.

Although Harvey credits Fabricius with the discovery of these valves, or portals, he goes on to say that:

> The finder out of these portals did not understand the use of them, nor others who have said the blood by its weight should fall downward: for there are in the jugular vein those that look downwards and do hinder the flow of blood upwards.

Then he describes what happens if you try to push a little rod into the veins:

> I have often tried in dissection if beginning at the root of the veins I did put the probe towards the small branches with all the skill I could, that it could not be driven by reason of the portals: On the contrary, if I did put it in outwardly from the branches towards the root, it passed very easily.

The most telling experiments, though, were carried out by tying a tight cord (a ligature) around one of the limbs, and monitoring what happens as it is released. Because the veins in the arm lie nearer to the surface than the arteries do, by adjusting how much the ligature is tightened it is possible to cut off the flow of blood along the veins alone, or along both the veins and the arteries. Harvey described how with a tight ligature around the arm the hand gets cold, but stays its usual colour. Slackening the ligature slightly allows blood to flow

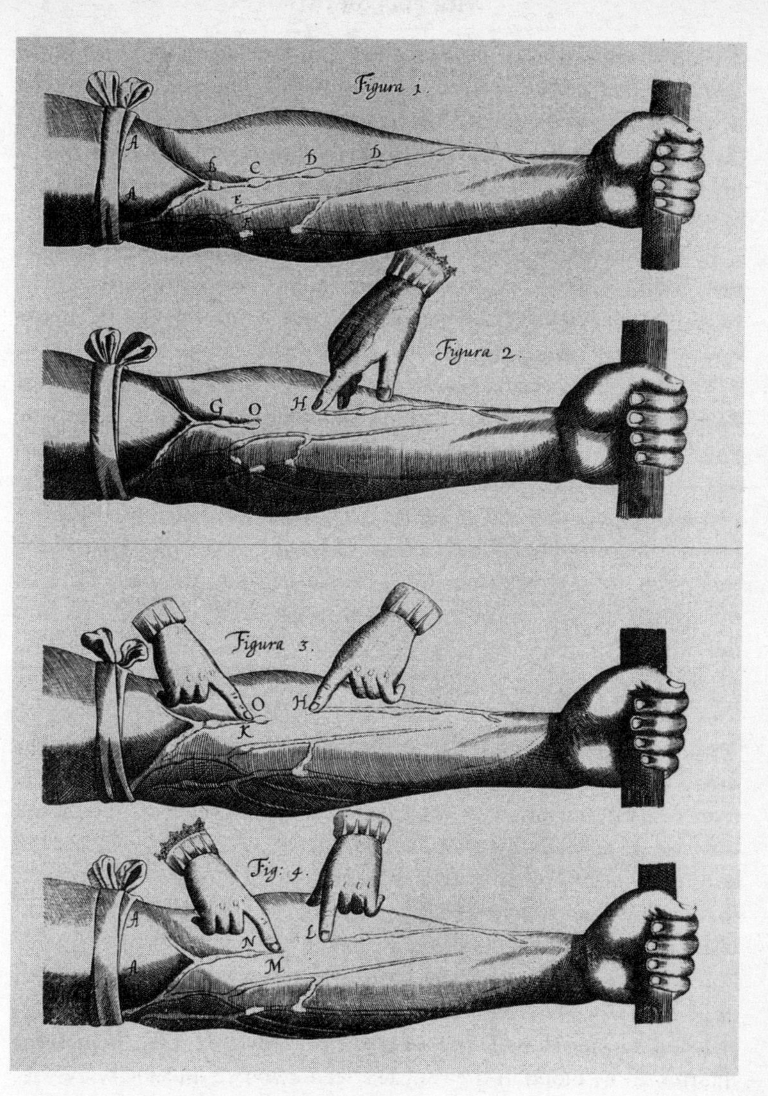


*From* De Motu, *circulation of the blood*

along the arteries into the hand, but not to escape along the veins. The hand flushes as the fresh arterial blood is pumped into it by the heart, and the veins on the 'upstream' side of the ligature (crucially, the side the hand lies on) swell up as the pressure builds. Finally, when the ligature is slackened the natural flow of blood is restored and the hand and arm resume their usual appearance and temperature.[16]

In an elaboration on this theme, Harvey described an experiment that could be carried out while the ligature was half-tight and the veins were distended. In this state, it is possible to see (and feel) little lumps in the veins, swellings which occur where two veins meet, or where the valves are located. If you put your finger just below one of these swellings and press it along the vein towards the hand, the swollen vein above the finger will disappear, as the blood is squeezed out of it, like toothpaste being squeezed out of a tube. Below the pressing finger, the vein becomes even more distended as the extra blood is forced into it – but only as far as the next valve. Even if the other hand of the experimenter is now used to do the same thing in the stretch of vein above the first valve, no blood gets through into the emptied stretch of vein. This powerfully demonstrates the function of the valves as one-way portals which only allow blood to flow *towards* the heart.

Harvey's second key evidence in favour of the circulation of the blood did not draw directly on experiments, but used evidence gained from dissection to show by logical reasoning that the blood could not possibly be continually manufactured in the liver (or anywhere else) and used up in the rest of the body. He started from the observation that during dissection he had found that the left ventricle of the human heart held a little more than 2 ounces of blood. He then estimated that the human heart must beat at least 1,000 times in half an hour (a deliberately low estimate, corresponding to a pulse rate of 33 beats per minute, about half that of a healthy adult at rest). If just one quarter of the blood in the ventricle were pumped into the body with every beat (again, a deliberately low estimate), then in half an hour 500 ounces of blood would pass from the heart into the body. These ounces (in the system of weights and measures used by apothecaries)

16. We would not advise you to try this, or any of the other experiments, at home!

are not the same as the ounces in the avoirdupois system, and there are just 12 of them in each apothecaries' pound, which itself corresponds to 373 grams. So in modern units 500 of Harvey's ounces is just under 16 kilos – very much a low estimate for the amount of blood seen to be leaving the heart, not just in one half-hour, but in every half-hour of life! Leaving aside the question of just how the liver could manufacture so much blood so quickly, Harvey pointed out that in order to make so much blood your body would surely have to consume at least the same amount of food (16 kilos) every half an hour, as well as disposing of the produce. Harvey also carried out similar calculations for other animals, and concluded that:

It is manifest that more blood is continually transmitted through the heart than either the food which we receive can furnish or is possible to be contained in the veins.

The only possible conclusion was that the *same* blood is continually being recycled, going round and round the body via the heart and lungs. Just as important as Harvey's conclusion, though, is his method. He didn't just say that 'obviously' more blood passed through the heart than the liver could produce; he measured the amount of blood in the heart and carried out quantitative calculations to make his case. This was where science differed from the natural philosophy of the Ancients. To quote Harvey again, from *De Motu*:

I do not profess to learn and teach Anatomy from the axioms of the Philosophers, but from Dissections and from the fabrick of Nature.

And yet, unlike Bacon, or Gilbert, or Galileo, Harvey saw himself as continuing in the Aristotelian tradition. This is most clearly seen in a passage from *De Motu* where he compares the circulation of the blood to the circulation of the Heavens and the circulation of water on Earth, each for its own mystical reasons, so that the motion of the blood is described as 'circular'

After the same manner that Aristotle sayes that the rain and the air do imitate the motion of the superior [Heavenly] bodies. For the earth being wet, evaporates by the heat of the Sun, and the vapours being rais'd aloft are condens'd and descend in showers, and wet the ground, and by this means here are generated, likewise, tempests, and the beginnings of meteors, from the circular motion of the Sun, and his approach and removal.

So in all likelihood it comes to pass in the body, that all the parts are nourished, cherished, and quickened with blood, which is warm, perfect, vaporous, full of spirit, and, that I may so say, alimentative; in the parts the blood is refrigerated, coagulated, and made as it were barren, from thence it returns to the heart, as to the fountain or dwelling house of the body, to recover its perfection, and there again by natural heat, powerfull and vehement, it is melted and is dispens'd again through the body from thence, being fraught with spirits, as with balsam, and that all the things do depend upon the motional pulsation of the heart.

So the heart is the beginning of life, the Sun of the Microcosm, as proportionably the Sun deserves to be call'd the heart of the world, by whose vertue, and pulsation, the blood is mov'd, perfected, made vegetable, and is defended from corruption and mattering; and this familiar household-god doth his duty to the whole body, by nourishing, and vegetating, being the foundation of life, and author of all.

So Harvey sees the heart as being the source of the warmth of the blood, the origin of the 'vertue' which makes living things alive; and although he is convinced of the *fact* of the circulation of the blood, he *also* subscribes to the Aristotelian idea of the perfection of circular (or at least, cyclic) motion. He saw the perfection of Heavenly motion carried over to the motion of air and water on Earth, and blood in the bodies of animals. But he did make the analogy between the action of the heart and a pump. In the page added to the Lumleian lecture notes around the time that *De Motu* was published, he writes:

It is certain from the structure of the heart that the blood is perpetually carried across through the lungs into the aorta as by two clacks of a water bellows to raise water. It is certain from the experiment of the ligature that there is a passage of the blood from the arteries to the veins. And for this reason it is certain that the perpetual motion of the blood in a circle is caused by the heart beat.

The 'clacks' Harvey refers to (which got their name from the sound they made in operation) are flaps of leather covering holes in the bellows/pump, and operating in just the same way as the valves in the veins – allowing liquid (in this case water) to flow easily one way through them but blocking the passage in the opposite direction. Later, in the second of two letters written in defence of his ideas to Jean Riolan, a physician at the Court in Paris, Harvey indicates that the kind of pump he is thinking about is the manually operated fire pump which could squirt a jet of water into the upper storey of a burning building. Making an analogy with the way blood from a severed artery jets out in bursts as the heart pumps, he says:

> Just as water, by the force and impulsion of a *sipho* [a common term at the time for a fire engine] is driven aloft through pipes of lead, we may observe and distinguish all the forcings of the Engine, even though it be a good way off, in the flux of the water when it passes out, the order, beginning, increase, end, and vehemency of every stroke.

But still Harvey thought that the heart was imbued with some mystic virtue or spirit, that was the source of life. It was left to René Descartes to put forward a purely mechanical model of the heart, part of his description of the body as a machine, with no need to invoke any life force or spirit. His rather bizarre notion envisaged stale blood entering the heart drop by drop, being evaporated, condensed in a re-vitalized form, and sent on its way again. This is an example of what happens when people try to build theories without doing experiments and making observations – anyone with experience of dissection and an observational knowledge of anatomy would have known, even in Descartes' day, that the idea was ludicrous. The difference between Descartes (and Bacon) and Harvey is that they talked the talk, while he walked the walk.

Harvey's ideas about the action of the heart and the circulation of the blood found some adherents immediately, but took a long time to become universally accepted. This wasn't just because of the existence of unreconstructed Aristotelians, who never grasped the importance of the experimental method and had to die off before science could become properly established. Even many of those of his contemporaries who were disposed to accept the evidence he presented, and even Harvey himself, were aware that one important piece of the jigsaw puzzle was missing. In 1628, and for several decades thereafter, there was no direct, observational evidence of the connections between arteries and veins which had to be there if blood really did flow out to the extremities of the body along arteries and back to the heart along veins.

At first sight, you might think that this left Harvey in the same position as those people who argued that blood must seep through the wall between the two ventricles of the heart, even though no openings were visible to the naked eye. But this is not the case. Harvey himself carried out experiments on animal hearts in which one chamber was pumped full of water under sufficient pressure to make it swell up, but without a single drop getting through the dividing wall into the other half of the heart. And the ligature experiment clearly shows that blood is going down the arteries into the hand, and back up the veins from the hand. The problem was simply that in the 1620s the existing technology – microscope technology – was not good enough to reveal the existence of the tiny capillaries linking the arteries and the veins in the extremities (and, indeed, throughout the body). Throughout the course of scientific history, progress in science has gone hand in hand with improvements in technology, in a mutually beneficial feedback. In this case, the improved technology of microscopes was used to reveal the tiny connections between arteries and veins by Marcello Malpighi, who worked at various universities in Italy from 1653 until his death in 1694. He made his great discovery in the early 1660s. And in a clear indication of the transition of the centre of scientific gravity from Italy to England since the middle of the sixteenth century, from 1667 onwards a great deal of Malpighi's work, carried out in Italy where the Renaissance had begun, was sent for publication to the Royal Society, in London. Even more symbolically, Malpighi was the first Italian to be elected as a Fellow of the Royal Society, in 1669. It is time to see how the Royal reached such a position of scientific prominence, starting out from meetings among natural philosophers held in Oxford during the Civil War and London a little later, in the space of little more than a dozen years.

# BOOK TWO

# Getting Started

*Dr John Wilkins*

# 4

# The Roots of the Royal I: The Oxford Connection

The origins of the Royal Society lay with a group of scientists who started holding regular meetings in Oxford at the end of the 1640s, during the Parliamentary Interregnum between the reigns of Charles I and Charles II. Although this was some ten years before William Harvey died, the Oxford group began to meet just after he had returned to London (no coincidence, since many of them were swept to Oxford by the same political winds that had swept him away), so Harvey was not directly involved in any of this, although his work was one of the examples that inspired them.

Some accounts suggest that the roots of the Royal go back even further, to meetings held in London earlier in the 1640s. Even some respectable historians subscribed to this view until quite recently, but it is now clear that although some of the same people who founded the Royal did attend those meetings, they were of a quite different character, and in no sense the forerunners of the Royal itself.

The confusion arose largely because of one man, John Wallis, who wrote a self-serving account of the early days of the Royal in 1678. Wallis, a Fellow of the Royal Society, was in dispute with another Fellow, William Holder, about priority in a scientific experiment. The details need not concern us, but Wallis essentially duplicated Holder's work, and published his results without crediting Holder. Holder (who was a brother-in-law of Christopher Wren) naturally objected to this, and published his account of the events. In establishing the chronological sequence of the experiments in that account, Holder referred in passing to the origins of the Royal:

Some years immediately before His Majesties happy Restauration, divers ingenious persons in *Oxford* used to meet at the Lodgings of that excellent Person, and zealous promoter of Learning, the late Bishop of *Chester*, Dr *Wilkins*, then Warden of *Wadham* College, where they diligently conferred about Researches and Experiments in Nature, and indeed laid the first Ground and Foundation of the Royal Society.[1]

Responding to Holder's criticism, Wallis clearly decided that attack was the best form of defence, and contradicted everything that Holder had said, including (seemingly just for the hell of it) his statements about the origins of the Royal. He said that the Royal had its roots in meetings held in London in 1645. None of this cut much ice with his contemporaries, who knew Wallis to be a clever man, but also both a liar and a plagiarist. John Aubrey, author of *Brief Lives*, sums him up like this:

Tis certaine that he is a person of reall worth and may stand with much glory upon his owne basis, needing not [to] be beholding to any man for fame, of which he is so extremely greedy, that he steales flowers from others to adorne his own cap, – e.g. he lies at watch, at Sir Christopher Wren's discourse, Mr Robert Hooke's, Dr William Holder, &c.; putts downe their notions in his note book, and then prints it, without owning the authors. This frequently, of which they complaine.

Through an accident of history, perhaps encouraged by a natural desire to push back the origins of the Royal as far as possible, later generations (who didn't know Wallis as well as Aubrey did) were taken in by Wallis's account. The details of how this happened can be found in Margery Purver's book, *The Royal Society*. It might seem that whether the Royal had its origins in London in 1645 or in Oxford in 1648 is a matter of hair-splitting of interest only to academics. But it really matters, and we have given this brief account here, because there was a key difference between the meetings in London and the meetings in Oxford, and it is that difference which made the Royal what it was, and which made the revolution in science.

The key point is that the meetings in London (which have been described by several writers, including Wallis himself) consisted of a

1. *A Supplement to the Philosophical Transactions*, Royal Society, London, 1678.

group of gentlemen who were interested in the new discoveries being made, and who met to discuss things like Fabricius' discovery of the valves of the veins, Harvey's ideas about the circulation of the blood, Gilbert's work on magnetism, Toricelli's invention of the barometer, and the ideas of Copernicus, Kepler, and Galileo. But, crucially, they carried out no new experiments themselves. The Oxford group, by contrast, actually called themselves an 'experimental science club'. They carried out new experiments, reporting the results back to the club. Their whole approach was a conscious rejection of the old Aristotelian way of teaching, and they specifically embraced the scientific philosophy of Francis Bacon – or at least, their interpretation of the philosophy of Francis Bacon. And the prime mover in all of this, as Holder correctly acknowledged, was the then Warden of Wadham College, John Wilkins.

The surprising thing is that there was ever any doubt about where and when the roots of the Royal became established. Almost as soon as the Society received its Royal Charter, it commissioned Thomas Sprat to write an official account of its origins, which was published in 1667. As we shall see, the founders of the Royal were well aware that they were attempting something revolutionary, and saw themselves as beginning the Baconian quest to bring nature within the understanding and control of humankind. They took great pains to ensure that posterity would know exactly how and why they had set out on this ambitious task, and, as Purver has established, Sprat's work was scrutinized by the Fellows, checked, re-checked and corrected, before it was allowed to see print. Sprat himself had been an undergraduate at Wadham College, and was a protégé of John Wilkins; he later went on to become Bishop of Rochester.

So why was the importance of Sprat's book largely unappreciated from the end of the seventeenth century to the second half of the twentieth century? The blame seems to lie with Thomas Birch, who was Secretary of the Royal Society in the middle of the eighteenth century, and published the Society's early minute-books, up to the end of 1687, in a volume titled *The History of the Royal Society of London*. Birch preferred Wallis's account, probably simply because it gave the Royal a slightly longer pedigree, and, following Birch, for two hundred years historians treated Sprat's account as the unreliable work of an individual writer who was ignorant of the London meetings. After an edition printed in 1734, Sprat's *History of the Royal Society* wasn't even published again until 1959. It was only around that time that it began to be appreciated, for the first time in nearly three hundred years, just how 'official' and reliable Sprat's account was. A salutary lesson in the importance of going back to original sources wherever possible, and not relying on what 'everybody knows' to be the truth – a lesson as important in science as in history.

So how did those meetings in Oxford get started? And where did they lead? The men who became the founders of the Royal Society were thrown together, separated, and re-arranged in different patterns, like the pieces of coloured glass in a kaleidoscope, by the turmoil of the Civil War and its aftermath. The pattern they eventually settled in, after key members of the group migrated from Oxford to London, centred around Gresham College, in London, which we mentioned earlier.

The college, established in 1596, was housed in what had been the mansion of Sir Thomas Gresham, an extremely wealthy merchant who had been financial adviser to each of the three children of Henry VIII who ascended the throne, who founded the Royal Exchange as London's business centre, and almost by the by set up the first paper mill in England. He was married to a sister of the first wife of Sir Nicholas Bacon, and in the small world of Elizabethan society knew his sons, including Francis Bacon, although long before Bacon was known as a philosopher of science. The site of the building that became Gresham College, in the City of London, is now occupied by the skyscraper known as Tower 42 (formerly the NatWest Tower).

Gresham's bequest provided funds to convert the house into a college, providing accommodation for seven professors at a time, and an income for them in perpetuity. The professors were to be distinguished experts in law, physic (medicine), rhetoric, music, chemistry, and astronomy, and their only duties were to give weekly lectures in term time, in both Latin and English, so that both visiting scholars from abroad and the educated gentlemen of London could benefit from their knowledge. The professors were also required to be celibate, although, as we shall see, this was interpreted in some cases in its literal sense of being unmarried, rather than implying abstinence from sexual activity.

Gresham College still exists, in different premises, although its importance has waxed and waned over the years. Its chief claim to fame is that it was indeed the focal point of the meetings in London that led to the formation of the Royal Society, and that it provided a home for the Society in its early years. Indeed, the inception of what became the Royal Society can be dated precisely, to a meeting of twelve men at Gresham College on Wednesday 28 November 1660, following a lecture there. They were the core of a group of people interested in science who had got into the habit of meeting up once a week or so after these lectures; but now they decided the time had come to do something more. Their twelve names are recorded in the Journal Book of the Royal Society as 'The Lord Brouncker, Mr Boyle, Mr Bruce, Sir Robert Moray, Sir Paul Neile, Dr Wilkins, Dr Goddard, Dr Petty, Mr Ball, Mr Rooke, Mr Wren, Mr Hil', and we are told that on that day they constituted themselves into an association 'for the promoting of Experimentall Philosophy'. This was obviously no spur of the moment decision, since proper notes were kept of the meeting. 'It was agreed,' the Journal Book continues,

> That this Company would continue their weekly meeting on Wednesday, at 3 of the clock in the terme time, at Mr Rooke's chamber at Gresham College; in the vacation at Mr Ball's chamber in the Temple, and towards the defraying of occasional expenses, every one should, at his first admission, pay downe ten shillings and besides engage to pay one shilling weekly, whether present or absent, whilst he shall please to keep his relation to this Company. At this Meeting Dr Wilkins was appointed to the chaire, Mr Ball to be Treasurer, and Mr Croone, though absent, was named the Registrar.
>
> And to the end that they might be the better enabled to make a conjecture of how many the elected number of this Society should consist, therefore it was desired that a list might be taken of the names of such persons as were known to those present, whom they judged willing and fit to joyne with them in their designe, who, if they should desire it, might be admitted before any other.

The subscriptions indicate both the wealth of these men and their commitment to the cause. It is impossible to make a direct comparison, but in round numbers, 10 shillings then corresponds to about £500 today. It sounds a lot, and it is, but many people today would happily pay a £500 initial fee and £50 a week for a season ticket to watch a Premier League football club. Another way of looking at it is that the pay for a university professor at the time might have been about £1 or £1 10s a week, so the subscription would represent one-twentieth of that salary.

One of the most significant features of the establishment of this society for the promoting of 'Experimentall Philosophy' was the election of Dr Wilkins to the chair. This is the same Dr Wilkins referred to by Holder as being, a few years earlier, the Warden of Wadham College in Oxford; it is clear that he was the leading scientific influence in bringing together, first, the group in Oxford and then, after the migration south, the group that gathered around Gresham College. For political reasons, he later moved into the background as the association of experimental philosophers became the Royal Society. But to understand the background to that meeting at Gresham College on 28 November 1660, it is best to look at the background of the twelve persons present, and a few of the other founders of the Society who happened to be absent that day, like Mr Croone; pride of place, of course, goes to the Right Reverend John Wilkins.

Wilkins came from an Oxford family (his father was a goldsmith), but he was born in Northamptonshire, at the home of his maternal grandparents (his grandfather was a vicar), in 1614. He was educated in Oxford, entered the university at the age of 13, and received his BA in 1631 before taking holy orders. His first real job, after a spell working as a tutor in Oxford, was as vicar of his grandfather's old parish. But this didn't last long. Wilkins was ambitious, and managed to get appointed as personal chaplain to a succession of noble gentlemen, which eventually involved moving to London. He became known both for his ability and as a supporter of the Parliamentary side in the Civil War, and when Royalist sympathizers were ejected from their posts in Oxford at the end of the war, in 1648 Wilkins was appointed as Warden of Wadham College.

These were not easy times for the university, even for supporters of Parliament. There were factions that wanted to do away with the elitist system altogether, and replace it with something more egalitarian. But in 1656 Wilkins found a way to cement his position. Robina, the sister of Oliver Cromwell (who was by now Lord Protector), became a widow when her husband Peter French, who had been Canon of Christ Church, Oxford, died. She was 62, and had nowhere to live, since her husband's accommodation went with the job. Wilkins was twenty years her junior (indeed, Robina had a daughter, Elizabeth, almost the same age as Wilkins), and besides, it was a stipulation of his post as Warden of Wadham that the holder be unmarried. And yet, he obtained a special dispensation from the Lord Protector to marry Robina.

It's easy to see what she got out of the arrangement – security and comfort for the rest of her life. Cromwell got his sister off his hands. But what was in it for Wilkins? Was he just interested in making his own position, as the Lord Protector's brother-in-law, secure? John Evelyn (who was, incidentally, a close friend of Robert Boyle, even though he was seven years older than Boyle) suggests there was a more honourable motive. He wrote in his diary that Wilkins (whom he knew well) was

> A most obliging person, but had married the Protector's sister, to preserve the Universities from the ignorant Sacrilegious Commander and soldiers, who would fain have been at demolishing all bothe places [Oxford and Cambridge] and persons that pretended to learning.

Wilkins was certainly held in high esteem at both universities. In 1659, when the Mastership of Trinity College, Cambridge, fell vacant, the Fellows petitioned Parliament to appoint Wilkins to the post, and Parliament obliged. But just as Wilkins had gained his post at Wadham when the Royalist incumbent was ejected by Parliament, on the Restoration of Charles II he was ejected from the post at Trinity to make room for a Royalist. Which is how he came to be in London, where he was lodging with a friend, Seth Ward, at Gray's Inn, in November 1660.

Ward himself had been Savilian Professor of Astronomy in Oxford, and Master of Jesus College, but had fallen out with Cromwell and been ejected in 1657. He was very much part of Wilkins' scientific circle and although not a founder of the Royal Society he was on the list of likely people drawn up at the initial meeting, and was elected as a Fellow in December 1661, shortly before he became Bishop of Exeter. The setback to Wilkins which saw him lodging with Ward in 1660 seems to have caused only a temporary hiccup in his own ecclesiastical career. After several appointments in various parishes, he became Dean of Ripon and then, in 1668, Bishop of Chester. This was an indication of his complete rehabilitation; the post, which he held until he died, just four years later, at the age of 58, was an extremely important one, in the gift of the King himself.

Wilkins was a gregarious and friendly man who inspired respect and encouraged people from widely different backgrounds to come together to discuss scientific ideas and, crucially, to carry out their own scientific experiments. His tolerance was unusual, but by no means unique, in an age which, as we have seen, was torn by religious disagreements over ideology. As a liberal, he was criticized for, among other things, such dangerous ideas as reconciliation with the Dissenters, and it is significant that the Royal Society itself, when it was set up, specifically encouraged men from all parts of the Christian Church to be among its Fellows. This sounds fairly narrow today, but was extremely tolerant and broadminded for the 1660s.

The Society's liberal views on religion were matched by its aim of doing good for humankind as a whole, very much in the way Bacon had envisaged. These attitudes of religious toleration and inclusion, and working for the good of humankind, echo some of the basic tenets of Freemasonry, and some of the founders of the Royal Society were indeed Masons, which gave them an additional bond with each other and with Charles II, himself a Mason. But it is probably going too far to suggest, as some have, that the Society was a product of Masonry. Rather, men with liberal and charitable views would be likely to be attracted to the Freemasons, and the same men would be likely to be attracted to the idea of science in the service of humankind.

Wilkins' own scientific interests tended towards the mechanical, so he was more of an engineer than a theorist. He devised a machine to raise a fine mist in his gardens, to produce artificial rainbows, and constructed transparent beehives in which the activities of the bees could be observed. He was also something of a practical joker, and in the grounds of his house he had a hollow statue with a long concealed pipe connected to its mouth, so that by standing concealed in the bushes and speaking through the tube he could make it seem to his astonished visitors that the statue was speaking.

He was also interested in astronomy, and as early as 1638, when he was 24, published a speculation about the possibility of one day travelling to the Moon. Two years later (and still two years before the death of Galileo) he published an account of the Copernican model, and lent his weight to the argument that the Earth is a planet. In addition to all this, Wilkins was keenly interested in language. He tried to devise a universal written language, and encouraged the use of plain language in writing, something which became a feature of the Royal Society's style. A contemporary described him as

> A person endowed with rare gifts; he was a noted Theologist and Preacher, a curious Critick in several matters, an excellent Mathematician and Experimentist, and one as well seen in Mechanecismes and new Philosophy, of which he was a great Promoter, as any of his time. He also highly advanced the study and perfecting of Astronomy both at Oxford . . . and at London.[2]

Half of the men who attended the meeting on 28 November 1660 had Oxford connections and had come directly under the influence of Wilkins as a promoter of the 'new Philosophy'. The striking development that occurred in Oxford under Wilkins' influence was that those people worked together on joint projects, in a sense under his supervision. Science was no longer solely the province of isolated men of genius working on their own. In the most telling example of this approach, as we shall see, it was Wilkins who introduced Robert Hooke to Robert Boyle, and set them the task of investigating what happened under conditions of low atmospheric pressure. Wilkins himself was more like the head of a modern university department or a professor with a team working under him, than an individual such as Gilbert or Galileo, and it was this that made Oxford the centre of the scientific revolution. Assuming Wilkins realized what he was doing (and it is hard to believe he did not), there is a clear Baconian influence at work here.

2. Quoted by E. Bowen and H. Hartley, in *The Royal Society – Its Origins and Founders*, ed. H. Hartley.

If we take the members of Wilkins' group more or less in chronological order, going by the time when they began to take part in the meetings of scientifically interested people which led to the formation of the Royal Society, the next one to mention is Jonathan Goddard. He was a physician who was born in 1617, studied at both Oxford and Cambridge, and became a Doctor of Medicine in 1643. Alongside his medical practice, he took part in the discussion group which Wallis mentions in his attack on Holder.

Goddard was a confirmed Parliamentarian, and served Cromwell as physician-in-chief to his army on the campaigns in Ireland in 1648 and Scotland in 1649; his reward was the Wardenship of Merton College, in Oxford, which he received in 1651. He was appointed as Professor of Physic at Gresham College in 1655, but didn't move to London until 1658, and even then retained the Wardenship at Merton until he was dismissed by Charles II in July 1660. He was never a great scientist, but rather a distinguished physician who was interested in science, and who provided one of the links between the Royal Society, the Oxford group, and Gresham College – many of the early meetings of the Society were held in his rooms at the college. He died in 1675.

William Petty deserves a whole book to himself, and any brief summary of his life sounds more like the outline for a work of fiction than a true account of a real life. But he was real. Born in 1623 in the village of Romsey, in Hampshire, Petty went to sea in his early teens, was shipwrecked in Normandy in 1638, and stayed there for a while teaching English to one Frenchman, navigation to another, and studying at the Jesuit academy in Caen after more or less knocking on the door and talking his way in (in Latin, of course). With this adventure behind him, he joined the Royal Navy (not that this was much of a navy in the days before Pepys), but when the Civil War broke out he moved back to France, where he studied in Paris under Thomas Hobbes for a year; he also met René Descartes. He then moved to the Netherlands to study chemistry and medicine, finally returning to England and settling in London in 1648.

With the general move of Parliamentary appointees to replace displaced Royalists in Oxford, Petty became a Fellow of Brasenose College, completed his qualification as MD, and, still only 27 years old, became Professor of Anatomy there and then Vice-Principal of Brasenose in 1650. He was also awarded the Chair of Music at Gresham College, and like Goddard benefited from two sources of income even though clearly he could not carry out both sets of duties. In fact, for several years he didn't do either job. With official leave of absence, in 1651, like Goddard before him, he became Chief Physician to the army in Ireland. This gave him the opportunity to display another talent.

After suppressing the rebellion in Ireland, the Commonwealth army had seized the lands of the rebels and intended to distribute them among its supporters. This was a huge task, involving a new survey of virtually the whole of Ireland, which was completely bogged down when Petty arrived there. He coolly offered to take over the task, and oversaw everything from the manufacture of compass needles for the surveyors to the final printing of the resulting maps. Carrying out the project to a tight deadline on a fixed-price contract with the government (often a way to ensure bankruptcy!), Petty produced a survey which has remained the basis for establishing legal title to land in much of Ireland ever since – and he made a profit of £9,600 for himself in the process. This was the basis of what became a substantial fortune.

While he was in Ireland, Petty also renewed contact with Robert Boyle, of whom more shortly, whom he had met briefly in London. He acted as Boyle's physician on at least one occasion, they became friends, and Petty was one of the links between Boyle and Wilkins that eventually led to Boyle's move to Oxford. In a long and successful life (he died in 1687) Petty eventually went on to apply statistics to the study of populations and economics, becoming the founder of economic statistics; he designed ships; he supported the idea of a decimal system of weights and measures; and, about three hundred years ahead of his time, he proposed what amounted to a national health service. But all of that came after his return to England and the foundation of the Royal Society. Like so many Parliamentarians, Petty was stripped of most of his appointments on the Restoration, and had to fall back on his Chair of Music at Gresham College as a base – which is why he was at the 28 November meeting.

One of Aubrey's anecdotes sheds insight into both Petty's view of the world and the attitude of the Royal Society. The Annual General Meeting of the Society was held on St Andrew's Day, and at one of these meetings Aubrey joked to Petty that it might have been better to have chosen the feast of St George, rather than that of the patron saint of Scotland. 'No, said Sir William, I would rather have had it on St Thomas day, for he would not believe till he had seen and putt his fingers into the Holes.' This was part of the attitude that made the Royal, from the outset, more than a debating society – to believe nothing on hearsay, like Doubting Thomas, and to carry out experiments to test new ideas.

The name of Christopher Wren is familiar enough, but not usually in such a scientific context. It may come as a surprise to learn that in 1660 he was Professor of Astronomy at Gresham College, and that he had given the lecture on 28 November which preceded the meeting at which the basis of the Royal Society was established. He was also very much one of John Wilkins' protégés, and rather younger than most of the others present at the meeting. To give you some idea of the esteem in which he was held by his contemporaries, when Robert Hooke published his book *Micrographia* in 1665, he referred in the Preface to Wren's earlier work, and said that:

> The hazard of coming after Dr Wren did affright me; for of him I must affirm, that, since the time of Archimedes, there scarce ever met in one man, in so great a perfection, such a mechanical hand, and so philosophical a mind.

Wren was born in 1632, in the village of East Knoyle near Salisbury, in Wiltshire. His father, also called Christopher, was the rector there. Young Christopher's uncle, Matthew Wren, was a senior churchman with influence at Court. When he was a student in Cambridge, he had been taught by Francis Bacon's friend Lancelot Andrewes, and more recently he had accompanied the Duke of Buckingham and the Prince of Wales on their unsuccessful visit to Spain to attempt to arrange a marriage between Prince Charles and the Infanta Maria. In 1635, he was Dean of Windsor, but was promoted to become Bishop of Norwich, and arranged for his younger brother Christopher ('our' Christopher's father) to take his place. Matthew Wren had a son, also called Matthew.

This family tradition of the Wrens of naming sons after their fathers is even more confusing in the case of the Christopher Wrens, since Christopher Wren junior was actually the second son of his father; the first baby boy had been born in 1631, and also named Christopher, but had died shortly after he was born. Christopher's mother also died young, when Christopher was only two years old, having produced nine daughters, as well as the two boys, since her marriage in 1623. After his father was appointed as Dean of Windsor in 1635, the boy was brought up in the grounds of Windsor Castle, and among the Court of Charles I, where he was a playmate of the young Prince of Wales, later Charles II, who was just two years older than him. At the end of the 1630s, John Wilkins was also resident at Windsor, as the personal chaplain to Prince Charles Louis, the exiled Elector Palatine, who had sought refuge at the Court of his uncle, King Charles. Charles Louis and his younger brother Rupert, who would distinguish himself fighting on the King's side in the Civil War, were the sons of Charles's sister, Elizabeth. Wilkins became a friend of Christopher Wren senior, but it is hardly likely that he would have taken much notice of the Dean's young son.

In 1642, the Dean (along with other members of the Court) had to flee Windsor when it was attacked by Parliamentary forces. He moved first to Bristol, and then on to Bletchingham, in Oxfordshire, to live with the rector, William Holder, who had married Christopher's sister Susan. Possibly partly in order to keep Christopher away from the troubles of the time, his father sent him to Westminster School in London. Although Christopher Wren senior and John Wilkins were on opposite sides in the Royalist/Parliamentary conflict, they had a common bond as churchmen, and it seems that Wilkins helped to keep an eye on young Christopher, until Wilkins himself was appointed Warden of Wadham in 1648.

After leaving Westminster School in 1646, Wren spent most of his time in Bletchingham, where he was taught by Holder, until 1649; then Christopher, now 17, followed Wilkins to Wadham, where he came under the patronage of Wilkins and joined in the activities of the experimental philosophers, even though he was still an undergraduate. He took his BA in 1651, MA in 1653, but stayed in Oxford as a Fellow of All Souls College, indulging in scientific pursuits, until 1657, when he was appointed as Gresham Professor of Astronomy.

In his inaugural lecture at Gresham College, Wren made explicit reference to the influence of Baconian ideas on the thinking of himself and his colleagues, referring to mathematical demonstration, rather than philosophy, as 'the great *Organ Organon* of all infallible science'. But he identified William Gilbert as the first scientist. Referring to the shaking off of the tyranny of ancient ideas and the new freedom of the scientific investigation of the natural world, he said:

Among the honourable Assertors of this Liberty, I must reckon *Gilbert*, who having found an admirable Correspondence between his *Terella*, and the great *Magnet* of the Earth, thought, this Way, to determine this great Question, and spent his studies and Estate upon this Enquiry; by which *obiter*, he found out many admirable magnetical Experiments: This Man would I have adored, not only as the sole Inventor of Magneticks, a new Science to be added to the Bulk of Learning, but as the Father of the new Philosophy.

Wren also made a striking comment on what might be achieved as astronomical telescopes were improved:

A time would come, when men should be able to stretch out their eyes, as snails do, and extend them to *fifty* feet in length; by which means, they should be able to discover *two thousand* times as many stars as we can; and find the Galaxy to be myriads of them; and every nebulous star appearing as if it were the firmament of some other world, at an incomprehensible distance, buried in the vast abyss of intermundious vacuum.

This was still less than fifty years after Galileo published *The Starry Messenger*, and Wren is speculating that the stars are other suns, orbited by other earths.

In 1661 he returned to Oxford as Savilian Professor of Astronomy, but by then he had already played his part in founding the Royal Society, and he continued to have strong links with the Society, serving as its President from 1680 to 1682. He died in 1723, at the age of 90.

Wren got the job of Gresham Professor of Astronomy because the previous holder of the post, Laurence Rooke, had moved sideways into the Gresham Chair of Geometry. The reason seems to have been simply that the rooms that went with the Chair of Geometry were better than those of the Professor of Astronomy, and provided a more convenient meeting place for those gentlemen who stayed on after the lectures to discuss scientific matters.

Rooke was another of those gentlemen who had made the move from Wadham to Gresham College. And, as we have seen, the meeting at which the as yet unnamed Royal Society was established took place in those rooms at Gresham. Not much is known about his early life, but he was born at Deptford in Kent in 1622, and studied at Eton before going up to Cambridge in 1639. He suffered from poor health, and had to go back to Kent in 1643 to recover; his degree was awarded by proxy in February that year. But by 1650 Rooke was well enough to move to Wadham College to work with Wilkins and the Savilian Professor of Astronomy, Seth Ward. He also worked with Robert Boyle on chemical investigations, before taking up the Chair of Astronomy at Gresham College in 1648. His main interest was in the use of astronomical techniques (especially lunar observations) to determine longitude at sea – one of the most pressing practical problems of the day. But he did not live long enough to make the contribution to science that he might have done. In 1662 his poor health finally caught up with him when he contracted a chill, developed a fever, and died at the age of 40. Because of this, even though Rooke was a founder member of the Royal Society, he never actually became a Fellow; he died just before the Society's first charter was granted.

It should, incidentally, be no surprise that many of the founders of the Royal Society were interested in problems of ship design, navigation, tides, and weather. In Britain in the second half of the seventeenth century this was cutting-edge technology, comparable to the aerospace industry in the United States in the second half of the twentieth century. The surprise would have been if any scientist in Britain in the 1660s worth his salt hadn't been interested in nautical matters.

Another thing about many of the founders of the Royal Society, especially the ones we have mentioned so far, that looks odd at first sight is that so many of them were supporters of the Parliamentary side in the Civil War, men who had received advancement under Cromwell but who lost their posts in Oxford on the Restoration. Why were they so keen that their society should be 'Royal', and should have a charter from the newly restored Charles II? But you have to remember that by 1660 most people in England were sick of strife (that's why the King was restored in the first place) and were ready to settle for a quiet life, and a period of stability. The men who founded the Royal Society were pragmatists (John Wilkins' marriage to Robina French testifies to that) who wanted their Society to be as official as possible, whoever might be in power. Their attitude is surely best summed up by the words used by Edmond Halley, himself a prominent early Fellow of the Royal, during the period of political turbulence caused in the 1680s when the Catholic James II succeeded his brother Charles II, and was then replaced by the Protestant combination of William and Mary:

> For my part, I am for the King in possession. If I am protected, I am content. I am sure we pay dear enough for our Protection, & why should we not have the Benefit of it?

This kind of pragmatic approach to the political realities surely explains also why the first President of the Royal Society was not John Wilkins, who was the driving force behind its foundation, or a great scientist like Robert Boyle, but rather William, Viscount Brouncker. Brouncker was a good but not brilliant mathematician, who as well as being one of the Oxford set had the merit of having remained a loyal Royalist throughout the Interregnum, served as an MP in the Parliament which called Charles II back to England, and was, hardly surprisingly, on good terms with the King – the 'King in possession', as Halley would later put it.

William Brouncker was born some time around 1620. He was the son of Sir William Brouncker, one of the gentlemen who attended Charles I (his title was 'gentleman of the privy chamber'); he was also Vice-Chamberlain to the Prince of Wales, who would become Charles II, and must have known William Harvey. Sir William was created a viscount as a reward for his services to the King and Prince in September 1645, and died just two months later, when his son inherited the title at about the age of 25.

By then, the new Viscount Brouncker was in the middle of his medical studies in Oxford, which culminated in the award of the degree of Doctor of Physick in 1647. But although he had originally intended to become a physician, while he had been at Oxford he had become fascinated by mathematics, and gained a reputation in the field through his correspondence with people like John Wallis. Like many good mathematicians, Brouncker was also interested in music. While keeping his head down and avoiding getting into trouble for his Royalist leanings, he translated a paper about music written by Descartes into English, as well as carrying out his best mathematical work. It was natural that he should be one of the regular attendees at the Gresham College lectures and the after-lecture meetings that began around the time of the Restoration.

In the months after the meeting of 28 November 1660, Brouncker acted as a go-between between the fledgling Society and the King, who had known his father so well and was favourably disposed towards the second Viscount Brouncker. In 1662, Brouncker was appointed Chancellor to the Queen, Catherine, and made Keeper of her Great Seal; when the King granted the Society a Royal Charter, it included the declaration that 'we have assigned, nominated and constituted, and made . . . our very well-beloved and trusty William, Viscount Brouncker to be and become the first and present president of the Royal Society'. This was clearly just what the Society wanted, and Brouncker held the post for fifteen years, carrying out his duties with a zealous devotion, although he had by now virtually stopped his mathematical work.

Away from the Royal, Brouncker was one of the King's ablest servants. He made his particular mark as a member of the Navy Board, presided over by James, Duke of York (the King's brother). Samuel Pepys described him as 'a very able person' and 'the best man of them all' on the Board. It was largely thanks to the close cooperation between Brouncker and Pepys, who was Clerk to the Navy Board and later Secretary to the Admiralty, that the Royal Navy was put on a secure footing and became the fighting force that enabled Britain to become a great power. When Brouncker died in 1684, Pepys was the executor of his will; later that same year, Pepys himself became President of the Royal Society.

Back in 1660, though, the one truly great scientist present at the foundation meeting of the Society was Robert Boyle – assuming it was Robert Boyle who was present, and not his nephew Richard. Some confusion has arisen about this, because the Society's records tell us that Robert Boyle was one of the people invited by letter to join the

Society after that meeting. Either Richard Boyle was present and Robert was later invited to join, or Robert Boyle was present, and Richard was later invited to join, but the wrong name was recorded. The second option is usually regarded as the more likely, since there is other evidence that Robert Boyle was in London in November 1660. If he had been in London, it is hardly likely that he would have stayed away from the meeting, as even the briefest glance at his background shows.

Boyle came from another of the upwardly mobile families of the Elizabethan era – in the case of his father, another Richard Boyle, the upward mobility had taken him just about as far as it was possible to go. He was born in 1566, and came from a respectable but not particularly distinguished family. He attended the King's School in Canterbury at the same time as Christopher Marlowe, and just before William Harvey, intending to go on to a career in law. But both his parents died before he was 20, and Richard was left all but penniless. He decided to go to Ireland to make his fortune, which he did to remarkably good effect. By marrying a rich widow and wheeling and dealing (not always entirely legally) with her money, he got rich. When his first wife conveniently died, he got married again, this time to the daughter of the Secretary of State for Ireland, and became respectable. As his wealth increased still further, he was able to buy the title of Earl of Cork by greasing the appropriate palms, and eventually became possibly the richest man in either England or Ireland.

Robert Boyle (strictly speaking, the Honourable Robert Boyle, as the son of an earl) was a late addition to the family. Thirteen siblings (seven girls and six boys) preceded him before he was born on 25 January 1627, when his father was 60 and his mother 40. Just one more child, a girl called Margaret, came three years after him; complications connected with the birth killed their 43-year-old mother.

Robert was brought up mostly by family retainers, and educated at first by private tutors. But he later went to Eton, where he stood out as a talented scholar, and lodged for a time with the headmaster, John Harrison, who gave him the free run of his own library, which included books by Regiomontanus, the *De Revolutionibus* of Copernicus and William Gilbert's *De Magnete*. In 1639, when he was 12,

*Robert Boyle*

Robert and his brother Francis (who was then 15) were sent on the continental Grand Tour, accompanied by a tutor. The party was actually in Florence when Galileo died in January 1642. The resulting commotion about the loss of the great man encouraged Robert to develop his interest in science further. The tour should by then have been nearly over, but when rebellion broke out in Ireland (one of the precursors to the 'English' Civil War) Robert was told to stay away from the fighting, while the older Francis hurried home to help. By the time this phase of the Irish conflict had ended, in 1643, two of Robert's brothers (but not Francis) had been killed, and his father, the Earl of Cork, had lost control of his estates and most of his fortune; the Earl died in 1643, at the age of 76.

Robert Boyle returned to England in 1644 with no money and extremely limited prospects, and with the Civil War proper now raging. He had the good sense to avoid being seen to take sides in the conflict, and was able initially to find a home in London with his sister Katherine. She was by now Viscountess Ranelagh, but living apart from her husband. Katherine was thirteen years older than Robert, and was the nearest he had to a mother; she looked after him for much of his later life.[3] She was also sympathetic to the Parliamentary cause, and had many influential friends.

One of the few possessions of the late Earl which was still in family hands was a small estate at Stalbridge, in Dorset. He had left this to his youngest son, and, partly thanks to Katherine's connections, Robert was able to keep the property throughout the years of turmoil that lay ahead. He moved there in 1645, and studied theology (a lifelong interest) and chemistry, carrying out many experiments in his own laboratory. Boyle also corresponded with other scientifically minded people, and sometimes met up with them in London, often at meetings held in Katherine's house. He referred to this group as an 'invisible college', but nobody has ever quite been able to pin down exactly which scientifically minded people were part of the group. Almost certainly, though, there must have been considerable overlap with the group that John Wallis mentioned in his attack on Holder.

One of Robert's brothers, Lord Broghill, had played a prominent

3. She also employed John Milton as tutor to her own children.

part in crushing the Irish rebellion, and this put the family in good odour with Parliament when the political situation stabilized. So in the early 1650s Robert was able to go to Ireland on an extended visit to sort out the family interests and reclaim title to the estates. The desperate state of the Irish people shocked him, and at first he doubted his right to take an income from such an impoverished country. But in the end he decided that he could do more good as a benevolent landlord and by using part of his income for charitable ends, and he accepted what he saw as his duty. When everything was sorted out, his share of the income from the estates was £3,000 a year, more than enough for him to live comfortably and to give generously to charitable causes. But this wasn't the only reason for the next big change that now occurred in Boyle's life.

In 1653, while still in Ireland, Boyle was taken seriously ill (at least, he *thought* he was ill). He put himself in the hands of William Petty, who was by now doubling up as Chief Physician to the army in Ireland and as Surveyor General – he had the ultimate responsibility, among other things, for the surveying of the Boyle estates. Petty decided that there were three things wrong with Boyle. First, he was reading too much, 'which dulls the brain'; secondly, he was frightened of catching diseases, and a bit of a hypochondriac; thirdly, as a result of the hypochondria he was dosing himself with all kinds of unwise 'medicine'. Petty's candour seems to have gone down well with Boyle, and the two became firm friends. Petty taught Boyle anatomy and physiology, and how to do dissections.

Boyle seems to have met John Wilkins, now at Wadham College, in London before he left England; the connection was strengthened when Petty, impressed by his new friend's ability as an experimental natural philosopher, also told Wilkins all about him. Wilkins wrote to Boyle inviting him to join the group in Oxford, and to take up residence at Wadham. His interest in Boyle may have been as much for his financial clout as his scientific ability, since Wilkins was always on the lookout for rich men who could sponsor the scientific work. But in Boyle he got both a wealthy patron and a first-class scientist, albeit one who preferred to work in his own way. Boyle treasured his independence, and would later write proudly that he was 'never a Professor of Philosophy, nor a Gown-man'; but the thought of regular contact with like-minded people appealed to him, and as he had no need of support from anybody, in 1654 he moved to Oxford and took lodgings in a house next to the Three Tuns pub, on the High. Two of his nephews, Richard Boyle, the son of his eldest brother the second Earl of Cork, and Richard Jones, the son of his sister Lady Ranelagh, were also students there from 1656 to 1658.

Things worked out better than Wilkins could have hoped, since Boyle set up his own laboratory in Oxford, in addition to the one Wilkins had established, greatly increasing the scope for experiments of all kinds to be carried out. For many years he also subsidized the work of other members of the Oxford group, as well as carrying out his own work. In an autobiographical memoir that he wrote later in life, Boyle acknowledged how fortunate he had been in his birth. Being the heir to a great title and estates would, he said, have been 'but a glittering kind of slavery', while as a younger son he had the freedom to follow his own inclinations and the means to pay for what he wanted. 'A lower birth,' of course, 'would have too much exposed him to the inconveniences of a mean descent.'

In Oxford, Boyle went on to carry out important experiments in pneumatics and electricity, but he is best remembered today for his pioneering work as 'the father of chemistry', in which he tried to make alchemy scientific and thereby pointed the way towards modern chemistry. He wrote about chemistry in his influential book *The Sceptical Chymist*, published in 1661, and discussed the idea of atoms in *Origins of Forms and Qualities*, published in 1666. Boyle's books were among the first scientific writings to embody the principles laid down by Bacon that the story should be told without embellishment or flights of rhetoric, but as a straightforward account of what had actually been done in the experiments, what had been observed as a result, and what the theoretical implications were, including the implications for future experiments. It was a style that would be made famous by Isaac Newton, especially in his book *Opticks*, but which Boyle pioneered. Although he only moved to London in 1668, when he was 41, Boyle still played an important part in the early success of the Royal Society, attending meetings whenever he was staying in London, and even standing in as acting President for a few weeks in 1662. In his later life, however, he was more active in the promotion of Christian religion than in science. He lived with Katherine in London, and died there in 1691, just a week after his sister died.

But although Boyle was a towering figure in the scientific world of Restoration England, the greatest direct 'gift' that he made to the Royal Society was his former assistant in Oxford, Robert Hooke. Hooke is one of several Oxford men who would play a big part in the establishment of the Royal, and are worth mentioning here even though they were not present at the 28 November 1660 meeting. Indeed, Hooke himself was the most important person in making the Society a success in the 1660s, and thereby ensuring that it survived. John Wilkins was the guiding force in getting the people who became the Royal Society together, but Robert Hooke was the man who made it work. Without either of them, it would not have existed in the form it took. Because he is so important, we shall go into rather more detail about Hooke's life, both here and in Chapter Six, than we have about the lives of his fellow founders of the Society.

Hooke's background was very different from that of Robert Boyle. He was born at Freshwater, on the Isle of Wight, in 1635 – the youngest child of John Hooke, the curate of All Saints Church, and his wife Cecily. Although poor, John Hooke had the patronage of Sir John Oglander, a member of the most important family on the island. Like his patron, John Hooke was a High Anglican (as high as you could get without being a Catholic), and this included reverence for the King as God's representative on Earth. Robert Hooke would grow up to share these convictions, and to hold all the varieties of Nonconformism in equal contempt.

But he nearly didn't grow up at all. Robert was a sickly baby and often ill as a child, so that his parents did not expect him to live long. According to his own account, written much later, for the first seven years of his life he lived entirely on milk and fruit, the only sustenance that did not make him sick. Robert's nearest sibling was his brother John, born in 1630; the five-year age gap and Robert's illness meant that he was brought up almost as an only child, indulged by parents who did not expect him to live, and taught at home by his father or (more often) left free to wander the countryside and sea shore as he wished. Among other things, it was on these wanderings that he first became intrigued by fossils, whose remains are found in abundance in that part of the Isle of Wight.

Although the island was staunchly Royalist, when the Civil War broke out the islanders prudently yielded to the Parliamentary forces without a fight, and came under their control. The Oglander patronage was now of little use to John Hooke, but he was able to keep his modest post (if the Parliamentarians had dismissed every Royalist on the island, hardly anyone would have been left), although it seems likely that his income dried up. This hardship hit the family when Robert was 7, and may explain why he was forced to adopt a cheaper 'normal' diet around that time; but apart from that, life carried on pretty much as usual for another five years before the Civil War had a dramatic impact on the island. It was also at this time that Robert began to receive a more formal education. He became a pupil of William Hopkins, a Royalist friend of John Hooke who was also under the patronage of Sir John Oglander, and was headmaster of the school at Newport. But this seems to have been an informal arrangement between friends, with no money changing hands and Hopkins teaching Hooke at home, rather than Hooke attending the school.

During those years, Hooke developed an aptitude for making things. He made a model sailing ship, about a metre long, complete not only with sails and rigging that enabled it to sail across the harbour, but with tiny guns that fired blanks. When he was given an old brass clock to play with, he took it to bits and copied the parts in wood so skilfully that his wooden clock actually worked. It was the forerunner of many superb scientific instruments that he would design and build in later life.

In November 1647, Charles I escaped from his detention by Parliamentary forces at Hampton Court. Most sources agree that he was allowed to escape by Parliament in the hope that he would go into exile in France; but instead he fled to the Isle of Wight, which he saw as a Royalist base where he could receive support from across the English Channel before returning to the offensive in England. But without reinforcements from the continent, the island would turn out to be not so much a base camp as a rather large prison. At first, the King was treated with the respect he expected, and took up residence in Carisbrooke Castle, where he held court and received visitors. All the Royalist gentlemen on the island went to pay homage to him, and the King later toured the island, being wined and dined by important families such as the Oglanders. It is tempting to speculate that John Hooke and his 13-year-old son Robert may have been among the loyal guests at one of these functions, especially since for a time Charles stayed in the house of William Hopkins, but there is no real evidence for this. With no prospect of receiving reinforcements, Charles's status soon changed from that of honoured guest to prisoner, in the charge of the Governor of Carisbrooke Castle, Colonel Robert Hammond; and when negotiations with Parliament finally broke down, late in 1648, the King was taken across the Solent to Hurst Castle, and then on to his ultimate fate. By then, Robert Hooke's life had been changed forever. Ironically, just when Parliament was at the height of its strength and the Royalists were at their lowest ebb, he was helped on his way by his father's Royalist connections.

John Hooke died early in October 1648, and was buried on 17 October. No details of his cause of death survive, but he had been ill for some time, and had already made careful plans with his friends and in his will for Robert's future; Robert may even have already left the island, in line with those plans, before his father died. In later life, Robert would tell a romantic story about how he left the Isle of Wight with £40 or £50 as his inheritance and started out in London as an apprentice to the Dutch painter Peter Lely. When he found that the smell of the paint made him ill, and decided that he could learn to paint perfectly well on his own, he gave this up and used his money to further his education instead. But the story should perhaps be taken with a pinch of salt. All we know for sure is that by the end of the year Hooke was installed at Westminster School, in London, and that this seems to have been exactly what his father had planned.

The actual beneficiary of the living where John Hooke had been a curate was the rector, Cardell Goodman. Goodman was, of course, a staunch Royalist (he eventually lost the living as a result, in 1651), and became a good friend of John Hooke, who named him as one of three executors of his will. Goodman had been a pupil at Westminster School, from where he had gone on to Christ Church College in Oxford before becoming a Fellow of St John's College and then moving on to the Isle of Wight, where the Freshwater living was at that time in the gift of St John's. He knew the headmaster of Westminster School, Dr Richard Busby (another Royalist), well, and there is no reasonable scope to doubt that Robert Hooke ended up at the school because he was introduced to Busby by Goodman in response to the dying wish of John Hooke.

Busby was a strong personality who was in charge of Westminster School from 1638 to 1695, stamping his mark on the place and demonstrating his character by holding on to his post throughout the Interregnum in spite of his known Royalist views. When he was required to subscribe to the National Covenant, he simply stayed away that day, claiming he was ill. Particularly during the troubled period of the Civil War and its aftermath, he demonstrated his flexibility (and got into trouble with the school authorities) by admitting more boys than the school could teach on a formal basis. Hooke was one of those special cases, who did not attend ordinary classes concentrating on the Classics; many of them were instead given the opportunity to learn technical skills, so that they would be able to find work. This separated them to some extent from the sons of gentlemen, who had no need to worry about a career. Busby also waived the fees for pupils who could not afford to pay, and some of them, including Hooke, lived in Busby's own house. Busby soon recognized Hooke's unusual talent, and the two remained friends long after Hooke left the school. John Locke, who was a fellow pupil at Westminster, also became a friend of Hooke for life; other contemporaries included John Dryden and, three years ahead of Hooke, Christopher Wren.

It was at Westminster that Hooke developed a pronounced stoop, a curvature of the spine that he later attributed to long hours spent bent over a lathe in the workshops. But although this remark gives an insight into how Hooke spent his time at the school – in the workshops rather than in the classroom – it was probably a coincidence that the bent back developed then. A condition now known as Scheurmann's kyphosis causes a curvature of the spine which develops in adolescence and is more common in boys than in girls. A contributory factor is a poor diet when young, and that fits with Hooke's background, although there may be a genetic cause as well. Although this specific diagnosis may not be correct, it seems likely that Hooke was suffering from some congenital problem, not simply the effects of working in a bent position, since his stoop continued to get worse long after he left Westminster.

But Hooke didn't spend all his time at Westminster in the workshop or studying academic subjects. He developed his drawing skills, and became a more than competent musician, particularly on the organ. This proved particularly useful when the time came for him to go up to Oxford in 1653, the year he was 18. Hooke had no money, and his only hope of a university education was to receive some kind of scholarship. Even then, he might have stood little chance of gaining a place at Oxford in normal times, but in the years following the Civil War there were fewer students than usual at Oxford, and the colleges were eager to make up the numbers. Westminster School had a special relationship with Christ Church in Oxford, and sent many of its scholars there. Busby managed to find a place for Hooke as a choral scholar, on the basis of his musical skills. The skills were real enough, but the best part of the deal was that since the puritanical government had done away with such fripperies as choirs at church services, Hooke got the scholarship without having to provide any choral services in return. He was also listed in the college records as the paid servant (servitor) of a 'Mr Goodman'. It was normal practice for a poor student to act as the servant of a richer one, but we know nothing about this Mr Goodman. Perhaps he was a relative of Cardell Goodman (who died in 1653), or possibly Goodman had left money for Hooke to be paid the servitor's wage without any duties to carry out. Either way, Hooke had just about enough to live on, but was in need of any other source of income he could find.

His first step towards employment outside the walls of Christ Church (indeed, outside the university altogether) was to become the paid laboratory assistant of a local doctor, Thomas Willis, who had his own chemical laboratory where he carried out experiments. Willis had been a contemporary of Busby in his student days at Christ Church, so we don't have to look far to find the connection that introduced Hooke to him. Willis had obtained his BA in 1639 and MA in 1642, then served as a soldier in the army of the King during the first phase of the Civil War. He qualified as a doctor in 1646, and although Oxford had by then fallen to the Parliamentary forces, he took up residence in the city and practised there for the next fourteen years. He was a Fellow of Christ Church until 1648, when he was ejected for refusing to formally acknowledge the legitimacy of the new regime. As an active member of the Oxford circle of experimental philosophers headed by John Wilkins, Willis provided Hooke with an entry into that group, although it seems likely that he also had an introduction to Wilkins from Busby, who would have recommended him as an ideal assistant for anyone engaged in practical scientific experiments.

As a loyal Royalist, Willis benefited from the Restoration, and in 1660 he became Sedleian Professor of Natural Philosophy in the University of Oxford. Although he was an early Fellow of the Royal Society and moved to London in 1666, he held on to the Oxford post, virtually a sinecure, until he died in 1675. Willis was an enthusiastic follower both of Harvey's ideas about the circulation of the blood and his methods, and made important contributions, based on experiments, to the understanding of the way blood flows to the brain. He also seems to have been the first person to guess the importance and source of what we now know as hormones. He wrote that 'as the blood pours forth something through the Spermatic Arteries to the Genitals, so also it receives a certain ferment from those parts through the veins; to wit certain particles imbued with a seminal tincture are carried back into the blood'. He said that those particles were responsible for the way an adolescent boy's voice breaks, the change in pattern of hair growth, and other changes associated with adolescence. But, of course, he had no idea what the 'particles' were.

Hooke's introduction to the group of experimental philosophers opened his eyes to a new way of looking at the world. He later described how at these gatherings in Oxford in the second half of the 1650s

> Divers Experiments were suggested, discours'd and try'd with various successes, tho' no other account was taken of them but what particular Persons perhaps did for the help of their own Memories; so that many excellent things have been lost, some few only by the kindness of the Authors have since been made publick.

The key word there being 'try'd'; the Oxford group were carrying out experiments to test hypotheses on a regular basis, and this is what sets them apart from other contemporary groups of scientifically inclined gentlemen who met to discuss the latest ideas without getting their hands dirty. Wilkins and Willis were largely responsible for getting the group to fund what became almost a private research institute, using rooms on university premises (wangled by Wilkins) to house laboratories, meeting rooms, and even telescopes for astronomical observations. The group was thriving when Robert Boyle arrived in Oxford, and he quickly ascertained that the 20-year-old Hooke was the most able of the several assistants employed by members of the group, and hired him as his own right-hand man. Hooke never completed his degree, having found something much better to do with his life (he was awarded an MA anyway in 1663, the year he became a Fellow of the Royal Society).

As Boyle's assistant, Hooke was not only paid by Boyle but lived in his house. Their relationship very quickly developed more into one of scientific collaborators and friends than master and servant, although Hooke was always careful to observe the social proprieties appropriate to their difference in rank. As a scientific researcher in his own right, something rather more than a technician working at Boyle's direction, but paid by Boyle, Hooke became the first full-time, professional scientist.

At this time, Boyle was particularly interested in atmospheric pressure and the nature of the vacuum, following on from the work of Galileo's pupil Evangelista Torricelli, who had invented the mercury barometer in 1644. In such a barometer, a glass tube sealed at one end is filled with mercury and then inverted with the open end below the surface in a dish of mercury. The level of mercury in the tube drops to about 30 inches above the surface of the mercury in the dish (where it is supported by the pressure of air on the mercury in the dish), leaving nothing at all (a vacuum) above it. For his experiments, Boyle needed a pump that could remove air from a glass container. The existing pumps were hopelessly inefficient, but Hooke invented an air pump that stood up to the task. It's a sign of Hooke's skill that even ten years later there were only half a dozen good air pumps in the world, and three of them had been made in Oxford to his design.

This was as significant an event in its day as the invention of the first 'atom smashing' machine (cyclotron) was in the twentieth century, opening up a whole new realm of scientific investigation. It enabled Boyle (and Hooke) to carry out many important experiments; later work led to the discovery of what is now known as Boyle's Law, that at a particular temperature the volume of a gas is inversely proportional to the pressure on it. In his own time, and for long afterwards, Boyle's work with Hooke on pneumatics was regarded as his greatest scientific achievement, with his contribution to chemistry only being given prominence much later.

The discoveries that Boyle and Hooke made using Hooke's air pump were described by Boyle in his book *New Experiments Physico-Mechanicall, Touching the Spring of the Air, and its Effects* (usually just referred to as *The Spring of the Air*), published in 1660. It is particularly significant that in this book Boyle credits Hooke, by name, for his work on the air pump. Although he had many other paid assistants, Hooke is the only one ever to be credited in this way. By modern standards, the acknowledgement may seem less than Hooke's due, given the importance of his contribution to the collaboration; but by the standards of his day, which are the appropriate ones to use, Boyle was being generous in his acknowledgement of Hooke's contribution, and this demonstrates the high esteem in which Hooke was held by the Oxford group at the end of the 1650s, even if he was still only a paid assistant. His future as Boyle's scientific right-hand man, a member of the Boyle household with no need to worry about paying rent or where the next meal was coming from, satisfying work, and the recognition of other scientists, seemed secure. All that would change when Boyle 'gave' Hooke to the fledgling Royal Society to act as its first Curator of Experiments. Before we move on to that, though, there are still other members of the group that formed the Royal to introduce; and one of them, Henry Oldenburg, was also to some extent a gift from Boyle, although he would later turn out to be something of a thorn in Hooke's side.

Oldenburg was German, and came originally from Bremen, although his connection with the Royal Society was very much through the Oxford group. He was born in the middle of the second decade of the seventeenth century (the exact date is not recorded), so he was about 40 in 1655. His father taught at the city's Gymnasium (in those days, something rather more than a high school), and later became a professor at the University of Dorpat. With this background, young Henry had a good education, culminating in the award of the degree of Master of Theology in 1639. He then travelled to England and worked as a tutor for several aristocratic families until 1648. After four years touring on the continent, making contacts that would later prove invaluable to the Royal, Oldenburg returned to Bremen in 1652, and a year later he was appointed as the diplomatic representative of the city of Bremen in London, where Cromwell was now Lord Protector. His special brief was to ensure that England respected the neutrality of Bremen in the war between England and Holland, although he stayed on in the post after the war ended in 1654. In the course of this work, Oldenburg dealt with John Milton, who was at the time the Latin Secretary to Cromwell – an important position since Latin was still the international language of diplomacy. Their relationship grew from the formal into friendship, and in one of his letters to Oldenburg, Milton compliments the German on speaking English better than any other foreigner he had met.

Some time around the end of 1654, Oldenburg's diplomatic work must have come to an end, because in 1655 he became the tutor of Richard Jones, the son of Lady Ranelagh and nephew of Robert Boyle. This appointment seems to have resulted from his friendship with Milton, who had a few private pupils (including Richard Jones) even during his time as Latin Secretary, and was a good friend of Lady Ranelagh. So when Richard Jones and his cousin Richard Boyle went up to Oxford in 1656, Oldenburg was part of their entourage. Oldenburg also registered as a student at the university, but none of the party had any intention of actually taking a degree. Through Richard Jones, Oldenburg met Boyle, and through Boyle he soon became a visitor to meetings of the Oxford group of experimental philosophers, firing an interest in science which never left him.

After little more than a year in Oxford, Boyle's nephews left for the next part of their education as gentlemen, the Grand Tour of Europe. Oldenburg accompanied Jones on a tour that took in Switzerland, Germany, and France, culminating in a year spent in Paris before the party returned to England in April 1660. Throughout these travels

Oldenburg kept up an extensive correspondence with Boyle, which shows his continuing interest in science; although he was not present at the meeting on 28 November 1660, he was listed as one of the people to be asked to join the Society as founder members. When the Royal Society received its charter, Oldenburg was named, along with John Wilkins, as one of its two secretaries. He was interested in science, a friend of Boyle, could speak five languages (German, English, French, Italian, and Latin), and had made many important contacts on his foreign travels. Nobody could have been better equipped for the job.

Although Oldenburg continued to earn his living by teaching and translating, and received only modest financial assistance from the Society to cover his expenses, he was responsible for the day-to-day running of the Society, and summed up the Secretary's duties himself, writing in the third person that

> He attends constantly the meetings both of the Society and the Councill; noteth the observables said and done there; takes care to have them entered in the Journal- and Register-Books; reads over and corrects all entrys . . . writes all letters abroad and answers the returns made to them, entertaining a correspondence with at least 50 persons; employs a great deal of time and takes much pains in satisfying forran [foreign] demands about philosophical matters . . .[4]

Oldenburg also started the first scientific journal in the world, the *Philosophical Transactions* of the Royal Society, and both edited and published it for the first twelve years of its existence, commencing in 1665. His extensive correspondence with foreigners got him into trouble in 1667, when he was arrested on suspicion of spying during the latest Dutch war. He spent several weeks as a prisoner in the Tower, but was released without charge when the war ended. Apart from that, his life after becoming Secretary was largely routine and uneventful, and he died in 1677. The one big failing in his time as Secretary was his feud with Hooke, which we shall discuss later.

With people like these among its founders, the Oxford connection alone would surely have been enough to ensure the success of the

4. Quoted by R. K. Bluhm, in Hartley.

Society, although perhaps without the other founders it would not have been a Royal Society. But the roots of the Royal spread even wider. Although the other founders were not exclusively Royalists, just as the Oxford men were not exclusively supporters of Parliament, the Royalist leanings of many of them were an important factor in ensuring the King's approval of the venture, and for convenience we can label them King's men.

# 5

# The Roots of the Royal II: The King's Men

By and large, the second group of founders of the Royal Society came on to the scientific scene later than the members of the Oxford group, and they included nobody of the scientific importance of Boyle or Hooke (or even Wren); but they played an important part in the administration of the Society, and establishing its place in the society of Restoration England.

The first important representative of the King's men to join the group of experimental philosophers that founded the Royal Society was Sir Paul Neile. He had a particular talent at grinding lenses and making telescopes, and was interested in astronomy. He also seems to have had a talent for keeping out of trouble, since although he was very much part of the royal circle, he lived quietly at White Waltham, near Maidenhead, during the Interregnum, keeping out of the public eye and returning to Court with the Restoration.

Neile had been born in Westminster in 1613, the son of the Archbishop of York, Richard Neile. He was educated at Cambridge, and went on to become one of the Gentlemen Ushers of the Privy Chamber to Charles I, being rewarded with a knighthood in 1633, when he was still only 20. He married Elizabeth Clark, the daughter of the Archdeacon of Durham; their eldest son, William, turned out to be a brilliant mathematician, but he died in 1670 at the age of 23. Sir Paul served as an MP for Ripon in 1640, in the so-called Short Parliament, but essentially disappeared from public life during the Civil War and the period of the Commonwealth. With independent means and no need to worry about making a living, he seems to have spent a lot of time making telescopes and using them to carry out observations; surviving correspondence shows that Christopher Wren visited White

*King Charles II*

Waltham to use Neile's telescope in 1655, and in 1658 Neile gave a 35-foot-long telescope to Gresham College, where it was later used by Hooke. Neile was also directly in touch with John Wilkins; John Evelyn records in his diary for 8 May 1656 that 'I went to visit Dr Wilkins at Whitehall, where I first met with Sir P: Neale famous for his optic glasses . . .' The great objects of astronomical interest to all these gentlemen (still less than fifty years after the publication of Galileo's book *The Starry Messenger*) were Jupiter and its moons, and the peculiar appearance of Saturn, which had not yet been explained in terms of the now famous rings.

With the Restoration, Neile became once again a Gentleman Usher of the Privy Chamber, and an invaluable link between Wilkins, Wren, and the rest of the original Oxford group and the King. It is no surprise that he should have been present with his long-established scientific friends at the foundation meeting of what became the Royal Society, and although he made no great scientific contributions in his own right, his skill at telescope making continued to come in handy, both when he gave a telescope as a present to the King and when he presented a 50-foot instrument to the Society itself in 1664. His name was often linked with that of Sir Robert Moray, whom we shall meet shortly, as a go-between taking messages from the King to the Society, or vice versa. For example, the minute-book of the Society records that on 16 October 1661

> Sir Robert Moray acquainted the Society that he and Sir Paul Neile had kissed the King's hand in the society's name; and he was desired by them to return their most humble thanks to his majesty for the reference, which he was pleased to grant of their petition; and for the favour and honour due them, of offering himself to be entered one of their society.

So Charles II duly became an FRS.

Neile served on the Council of the Society continuously until 1673, and from time to time over the following five years. His business acumen and contacts were immensely valuable in helping to keep the Society afloat financially during its difficult early years; Neile's own investments included participation in the original Hudson's Bay Company. After 1678, when he was 65, Neile took no active part in the running of the Society, and he died in 1686.

Another amateur astronomer who was drawn into Wilkins' circle of experimental philosophers as a result was William Balle, who became the first Treasurer of the Royal Society. His name is often recorded as Ball, but since he wrote it with an 'e' himself, when signing the Journal Book of the Society, we shall stick with the longer version. Balle's father, Sir Peter, was the Recorder of Exeter and served as Attorney General both to Queen Henrietta Maria (the wife of Charles I) and Queen Catherine (the wife of Charles II). There is very little information about William's early life, but he was probably born in 1627, and since he was in residence at the Temple for a time, he may originally have intended to become a lawyer, like his father. But his father's success and Balle's inheritance as the eldest son meant that he didn't really have to work at all, and could indulge his passion for science.

Balle first appears as a member of the group who founded the Royal Society when his observations of Saturn are described by John Wallis in a series of letters to the Dutch scientist Christiaan Huygens, commencing in 1656. It was Huygens, of course, who was the first person to realize that there are rings around Saturn, in 1655; it took some time for this to be widely accepted, and Huygens used Balle's observations and drawings in support of his idea. Balle's work is also referred to in a letter from Wren to Neile, in 1658; and as a further indication of Balle's status in scientific circles, when Huygens visited London in the spring of 1661, he was the guest of honour at a dinner at Balle's London house. The particular occasion for the dinner, on 1 May, was to celebrate the first anniversary of the acceptance by Parliament of the Declaration of Breda, the statement by Charles II which paved the way for the Restoration; since Charles had actually sailed for England from The Hague, it was particularly appropriate to have a distinguished visitor from the Netherlands at the dinner. But Huygens could have had his pick of dinners that night, and it is significant that he chose to accept Balle's invitation. Balle was no mean astronomer, and a more accomplished scientist than Neile. He regularly attended the gatherings at Gresham College which led up to the meeting on 28 November 1660, and it is no surprise that he was present that day – the surprise would have been if he had not been present at a lecture given by his fellow astronomer Christopher Wren.

At that foundation meeting, Balle was appointed as the first Treasurer of the Society. Unlike Wilkins' position as Chairman, this appointment was later confirmed in the charters that made the Society Royal. Clearly, it did no harm to have a member of a family well known to the King in such a post. But Balle held the post only until 1663, when he resigned. It is Balle's scrupulous record of payments of subscriptions to the new Society which provides the best evidence that it was indeed Robert Boyle, and not his nephew Richard, who attended the foundation meeting.

In addition to his administrative work for the Society, Balle was made its Curator of Magnetics, and carried out many magnetic experiments both for himself and as demonstrations for the other Fellows. This work went on much longer than his tenure of the Treasurership; although Balle moved to Devon at the end of 1664 to take over his father's estate, he continued his magnetic experiments and astronomical observations there, and still attended meetings of the Royal Society in London whenever he could. In October 1665, William Balle and his brother Peter made observations of Saturn using a 38-foot-long telescope, and reported their observations in a letter to Sir Robert Moray (another founder member of the Royal), who had them published in the *Philosophical Transactions*. The last time Balle attended a meeting of the Royal was on 9 November 1676, when he gave the Society two pieces of amber; but as late as 1680 he writes to Hooke to describe his magnetic observations and his intention of constructing 'a magnetic needle 10 feet in length, and another of 20 feet in length, in order to examine the variation of the directive virtue'. He died in 1690.

When Balle resigned as Treasurer of the Royal Society in 1663, his successor was another of the founding Fellows, Abraham Hill. Hill was an astute businessman and an ideal person for the job, although less obviously a King's man than the other founders discussed in this chapter. He made no notable contribution to science at all, but was an intelligent, well-read man with a comfortable fortune, much younger than most of the other founders, who was fascinated by science and happy to pay his dues to the Society for the privilege of learning at first hand about the new discoveries that were being made.

Hill's father, Richard, was a cordwainer (a member of the guild of shoemakers) who became a successful and prosperous merchant, an Alderman in the City of London, and eventually Master of the Cordwainers' Company. Abraham, his eldest son, was born in 1635, and was brought up to follow in his father's footsteps as a merchant, specializing in foreign trade. But as Thomas Sprat records, from 1658 onwards Hill was among the 'several eminent persons' who got into the habit of gathering at least once a week at Gresham College after the lectures given by Wren and Rooke.

Hill's contribution to the Society might never have gone further than this, but both his parents died early in 1660, shortly before he was 25, leaving him with a comfortable fortune. This gave him the freedom to indulge his interest in science to the best of his ability, which meant serving the Royal Society in various administrative positions between 1663 and 1721. The fact that Hill was available and willing to do these jobs must have been a big factor, but the fact that the Fellows consistently elected him to important positions within the Society indicates both his skill as an administrator and the esteem in which he was held, in spite of his relative youth when he started and the fact that he had had no formal university education (nor, indeed, anything more than a basic school education). Hill was a Member of Council of the Royal from 1663 to 1666 and again from 1672 to 1721 (although he surely cannot have been a very active member in his 86th year). He served as Treasurer from 1663 to 1666 and again from 1677 to 1699. And he was one of the secretaries of the Society from 1673 to 1675. In 1665, during his first spell as Treasurer, Hill was granted the patent rights on several inventions, all of which were actually the work of various other Fellows; the Royal had made the patent applications in a single group under Hill's name for administrative convenience and to reduce costs.

Hill's business and financial interests continued to develop outside his work for the Royal, and when John Tillotson became Archbishop of Canterbury in 1691, Hill was appointed as his Comptroller of Finance; he held the post until Tillotson died in 1694. Politically, Hill seems to have been more of a Republican than a Royalist, which probably explains why his considerable talents were not used by either Charles II or James II. But when William and Mary came to the throne in 1689, and Parliament had a greater say in the running of the country, he became more acceptable to the establishment. Hill became a Commissioner for Trade at the Board of Trade in 1696, when he was 61, and worked diligently at this post until the beginning of 1702, retiring from public life at the age of 66 when Mary's sister Anne became Queen. For most of this period at the Board of Trade, Hill had particular responsibility for matters relating to the American colonies of New England, Newfoundland, and New York – the job was far from being a sinecure. In his retirement, Hill lived at the manor of St John's at Sutton-at-Hone, in Kent, which he had purchased with his inheritance in 1660. He died there in 1722, three years short of his 90th birthday.

The last two of the dozen founding members of the Society present at the 28 November 1660 meeting were Sir Robert Moray and Alexander Bruce, who later became the second Earl of Kincardine. They were both in the Royalist camp (though Moray got there by a roundabout route), both Scottish, and good friends; but Moray played a much bigger part in the establishment of the Royal Society than Bruce did. Although he was a relative latecomer to the group, Moray (and to a lesser extent Bruce) made a crucial contribution by bringing to the mix a first-hand appreciation of the value of applied science to agriculture and business. Whereas Bacon had theorized about the benefits to mankind of science, Moray had seen science in action. He'd also done a lot more.

Robert Moray was born on 10 March, either in 1608 or 1609 – the record of his early life is very patchy and incomplete. His father was Sir Mungo Moray, of Craigie, in Perthshire; and he had a younger brother, who as Sir William Moray eventually became Master of the Works to Charles II. One of the few facts we know about Robert's early life is that when he was about 15 years old he visited the Culross estate of the Bruce family, and saw Sir George Bruce's underwater coal mine in the Firth of Forth. Access to the mine was through a shaft which had been sunk from an artificial island near the Fife shore, surrounded by water at high tide; the mine itself extended for a mile out to sea. The mine, and the machinery used (with the aid of horse power) to pump water out, seems to have fired Robert's interest in technology and the application of science to the improvement of mankind's lot.

But it took a long time for this interest to bear fruit. Moray was probably educated at St Andrews University (certainly somewhere in Scotland), before, as Aubrey records, he 'betook himself to military employment in the service of Louis XIII'. This was in the days of the Auld Alliance between Scotland and France, and it seems Moray joined a Scottish regiment (the Scots Guard) which took up arms as mercenaries with the French in 1633. We don't know the details of his career, but Moray clearly made an impact in Paris, where Cardinal Richelieu was at the height of his power and influence. Before the end of the 1630s, Richelieu, who was keeping a careful watch on the growing unrest in Britain which would soon lead to civil war, had recruited Moray as one of his agents.

According to the historian Patrick Gordon,[1] when Richelieu felt the need for a Scottish spy,

> Choosing forth a man fit for his purpose amongst a great many of the Scots gentry that haunted the French court (for the reason of the ancient league betwixt the French and them, they love always to breed themselves in France) he chooses forth one, Sir Robert Moray,[2] a man endowed with sundry rare qualities, and a very able man for the Cardinal's project.

This was at the time when the Scottish Covenanters were rebelling against Charles I. With a promotion to Lieutenant-Colonel in the Scots Guard, Moray was sent back to Scotland, under the cover of recruiting more Scots for the Guard, but reporting back to Richelieu and with a brief from the scheming Cardinal to make as much trouble as possible for the English King.

In 1640, Moray became Quartermaster-General of the Covenanters' army, and he was with them when they took Newcastle from the English. His duties included designing and overseeing the construction of fortifications, and it is clear that by this time he had a thorough grasp of engineering principles. It was around this time that Charles I was seen at his worst, making promises to the Scots that he had no intention of keeping, and making a bad situation worse. Perhaps significantly, Richelieu died in 1642, and Moray may have felt that

1. *Short Abridgement of Britane's Distemper*, Spalding Club, 1844.
2. In fact, Moray had not yet been knighted.

this released him from some of his obligations to the French. Early in 1643, far from trying to stir up more trouble he was acting as a go-between between the Covenanters and the King, and it was during this work as a negotiator that he was knighted by Charles, on 10 January that year in Oxford.

What looks at first sight like a bizarre honour to give to an enemy actually made sound political sense. Moray wasn't at all anti-Royalist, but wanted the King to be more tolerant of the Scottish Dissenters. At a personal level, Moray and Charles got on well together, and the King was impressed by Moray's ability as a negotiator. Charles needed someone who could negotiate not only with the Scots but also with the French, and knew Moray was well in with the French (though he obviously didn't know just how well in), and in that role Moray had to be clearly identified as having the trust of the King. How better to show this than to knight him? Moray was not only a skilful negotiator; he became friendly with both the King and the Prince of Wales, the future Charles II, with results that would play a big part in the early history of the Royal Society. In the short term, however, he returned to France (which was now embroiled in the Thirty Years War) where a further promotion to full Colonel showed how well he was succeeding in all his tasks. In November 1643, while fighting with the Scots Guard, Moray was taken prisoner by the Duke of Bavaria, and was out of circulation when the first phase of the English Civil War played itself out. But he was ransomed by the French in April 1645 for the handsome sum of £16,500. This indicates that the French still considered him a valuable asset, even though Richelieu had been dead for three years.

The seventeen months that Moray spent as a 'guest' of the Duke of Bavaria were not particularly harsh; as a gentleman and a senior officer he was allowed to live in some comfort, to read, and to communicate by letter with his friends. We know that one of the books that he asked for and was allowed to study during his captivity was a volume on magnetism; this is further evidence of his interest in science.

Moray's value to the French was still as an agent or negotiator who could help them to make the most of the troubled situation in England. While he had been a captive of the Bavarians, Louis XIII had died and been succeeded by the boy king Louis XIV, with his mother, Queen Anne, as regent. The timing of Moray's ransom was just right, from their point of view, since Charles I was defeated at the Battle of Naseby on 13 June 1645. A few months later, Cardinal Mazarin, who was now the Chief Minister in France, sent Moray as a negotiator to join a party of French ambassadors in talks with the Scottish representatives in London.

In the horribly complicated politics of the time, Charles, defeated by Parliament, was negotiating with the Scots for their support. During the protracted negotiations Charles moved from Oxford to Newark and then on to Newcastle (where he was visited by Harvey) for the protection afforded by the Scottish army (he was really their prisoner). Moray was closely involved in the tedious negotiations, and when they were clearly going nowhere at the end of 1646 he tried to arrange for the King to escape to continental Europe, but the plan came to nothing. Charles was handed over to the English, and the Scots went back to Scotland.

Moray returned to France. There, he became involved in negotiations to persuade the Prince of Wales, who was now in his late teens and had escaped to France after the Parliamentary triumph, to return to Scotland. After Charles I was executed in 1649, his son took the title Charles II. In 1650, he landed in Scotland, where he was crowned King on 1 January 1651 and led a rather raggle-taggle army in an unsuccessful invasion of England that ended in defeat at Worcester. Famously, Charles hid in an oak tree during his escape, and went back to exile in France, at the age of 21. Moray, who had returned to Scotland with Charles, stayed on there and became a Justice Clerk and Privy Councillor, and then a Lord of Session, all in 1651 – but he didn't enjoy these posts for long, since Cromwell soon subdued at least the Lowlands of Scotland.

While the Oxford group was becoming established under the guidance of John Wilkins, Moray was leading a more exciting life as an active Scottish Royalist during the period of Cromwell's complete authority over England. Soon after Charles II had been defeated at Worcester, Moray had married Sophia Lindsey, the sister of another prominent Lowland Scots Royalist, the Earl of Balcarres. But this didn't stop him taking an active part in organizing an uprising of Scottish Highlanders in support of Charles II. Both adventures ended badly. Sophia died in childbirth (the baby was stillborn) on 2 January 1653, at the age of 28; and the Highlanders were crushed by Cromwell at Loch Garry in July 1654. Moray himself was briefly imprisoned by the Scots following a false allegation that he was involved in a plot to assassinate Charles II, but Charles dismissed the evidence as a clear forgery. So in 1655, now 46 years old and with his military career over, Moray returned to France and spent the next five years leading the life of a private gentleman in exile. He spent a year in Bruges, then moved on to Maastricht, where he lived from July 1657 to September 1659.

Although Moray lived almost as a recluse for those two years, we have more documentary evidence about his life then than at any other time, because of the extensive correspondence that he kept up, in particular with his friend Alexander Bruce. A huge collection of these letters, known as the Kincardine papers, survives, and testifies to Moray's longstanding interest in science, which he was now able to put into practice by carrying out many experiments, in particular in chemistry. To put what we learn from this correspondence into context, we need first to introduce Alexander Bruce himself – although we can do little more than introduce him, since so few relevant records survive that he remains a shadowy figure.

Bruce was born in 1629, so he was about twenty years younger than Moray; the relationship between them seems to have had a flavour of master and pupil, with the older man giving advice to the younger, but we have no information about how they met or what contact they had with one another before 1657 (although, as we have mentioned, we do know from his letters that Moray visited Culross at least once in the 1620s, before Alexander was born). Indeed, apart from his approximate date of birth we know nothing at all about Bruce before the letters in the Kincardine archive were written. We do know that he came from a wealthy Scottish family with extensive coal mining and salt interests, near Culross in Fifeshire, built up by his grandfather and father. And Alexander Bruce had an older brother, Edward, who was made Earl of Kincardine by Charles I in 1647. Since Alexander Bruce was living in Bremen in 1657, we can infer that both brothers were supporters of the King, and that Alexander, like Moray, was living in exile during the Interregnum.

The letters from Moray show that Bruce was seriously ill (with 'the ague') while he was in Bremen, and it has been suggested[3] that one of the reasons why Moray wrote so many letters to him at such great length is that he was doing his best to provide the younger man with an interest to relieve the tedium of lying ill in a foreign land. More of the letters themselves in a moment.

At least partially recovered from his illness (although he continued to suffer poor health for some time), when news of the negotiations between Parliament and Charles II reached Bruce in 1660, he set off for The Hague, and he was among the King's party that returned to London for the Restoration. He lived at Charing Cross for about a year, but by the end of 1661 he was back in Culross. When his brother died unmarried in 1662, Alexander Bruce became the second Earl of Kincardine, and his responsibilities kept him in Scotland essentially for the rest of his life. He was made a Privy Councillor in 1666, and in 1667 became a Lord of Session and one of the King's Commissioners for the Government of Scotland. He died in 1680, but had ceased paying his subscription to the Royal before the end of the 1660s. Quite clearly, he was present at the meeting at Gresham College on 28 November 1660 simply because he happened to be living in London and was a friend of Robert Moray. Of all the founder members of the Royal Society, he had the least direct influence; but his existence is of immense value to historians in shedding light on the role of Moray himself.

Bruce's interest in technical matters was not so much in the abstract business of science, but in practical matters, related primarily to the use of coal, but also to the problem of developing accurate portable timepieces for navigation at sea. One of the early letters from Moray among the Kincardine papers, written in December 1657, informs Bruce about the new developments in making coke from coal:

> It is a kinde of baking of coales that you would not for a groat were known at home for it makes them last much longer and burn farr hotter than otherwise. I do not mean to tell you the trick of it unless you say you groan for it, before you see it, but there is another new trick on fire in Holland which will spare the Brewers I think 2 parts of 3 of their Coals. How do you like that?

3. By D. C. Martin, in Hartley.

As a mine owner and coal exporter, we can guess that Bruce didn't think very much of it!

From the context of these letters, it is also clear that Moray kept up an extensive correspondence with other scientifically minded men across Europe, but none of these letters survive. Moving on from practical matters of direct relevance to the Bruce mining and salt interests (such as techniques for pumping water out of coal mines), he wrote at length about chemistry (including his own experiments), horticulture, clocks, whaling, medicine, and music. In one letter he describes how at the time of writing he has seven stills operating on two fires, and in another describes his latest furnace, which 'will gar your eyn reel when you see it'. One of the key themes running through all this is the element of practicality, the application of science to improve the lot of mankind, and to establish, in Baconian terms, the dominance of man over nature. Moray was also one of the first Freemasons, and used to sign his letters with a pentagram that he referred to as his Masonic sign; the correspondence with Bruce continued into the 1660s, but at a much reduced rate after the Restoration, when both men had many other things to attend to.

In September 1659, Moray went to Paris to meet Charles II, and he was closely involved in the negotiations which led to the Restoration. But he did not return to England with the royal party in May 1660; for some months he stayed in Paris, keeping up a flow of letters in an effort to ensure the best possible treatment for Scotland under the new regime. He was, however, in London by August 1660, in good time to have become a regular attender at the Gresham College meetings before 28 November.

It may seem surprising that a relative newcomer to the group should have played such an important role in founding the Royal Society, even given his extremely close relationship with the King. But it seems to me that Moray acted as a catalyst, coming from outside the Oxford group but with a profound interest not just in abstract science but in the practical application of science. This is not to say that the Oxford/Gresham group were abstract thinkers interested only in 'pure' science; far from it. They were keenly interested in items as practical as improved farming machinery and as impractical (in the seventeenth century) as the possibility of flying to the Moon. They also had no aversion to making money, as the efforts of several members of the group to devise a practical means of timekeeping at sea testify. They were, however, to some extent academics, working in the university system of the day. Moray, by contrast, was a man of the world, who had seen first hand the coal mines and salt pans of the Bruce family on his visits to Culross, and had a vision of something like the industrial revolution that lay not so far in the future.

Without Moray, what had by then become the Gresham group might have bumbled along in much the same way for some time, and possibly even formalized its status as a society of some kind. But Moray seems to have been the man with the Big Idea, who inspired the group to see themselves as the people who could make something like Bacon's Foundation a reality. Combining this with his closeness to the King (Aubrey says that Moray 'had the king's eare as much as any one'), although John Wilkins was the most important person in bringing the group of scientists together, Robert Moray was the most important person in turning that group of scientists into the Royal Society. At the very first meeting after 28 November, on 5 December, the Journal Book of the as yet unnamed society records that 'Sir Robert Moray brought in word from the Court, that the King had been acquainted with the designe of this Meeting. And he did well approve of it, and would be ready to give encouragement to it.'

What happened after that is the subject of the next chapter. But although he was extremely active in the early years of the Society, carrying out experiments, serving on committees, and playing a major part in fund raising, Moray was still primarily a Scottish politician working to do his best for King and Country. He was re-appointed to his post of Privy Councillor in February 1661, and later that year became a Lord of the Exchequer, one of the inner group of courtiers that effectively ruled Scotland. In 1667 he was sent to Scotland as a member of a commission which introduced various political reforms, and on his return to London was greeted by the King 'with crushing and shaking [of his] hand, and with as good looks and as much kindness as I could wish'. In London, before and after his visit to Scotland, he lived in rooms at the Palace of Whitehall provided by the King, adjacent to the King's own chemical laboratory, where the two of them often carried out experiments together. On 15 January 1669 Samuel Pepys wrote in his diary:

> Then down with Lord Brouncker to Sir R. Murray, into the King's laboratory under his closet; a pretty place; and there saw a great many chymical glasses and things but understood none of them.

A year later, Moray retired from public life at the age of 62; he died on 4 July 1673, and was buried in Westminster Abbey. His importance to the Royal Society has been summed up by the historian D. C. Martin, writing on the occasion of the tercentenary of the Royal:

> That the charters and statutes of the Society, which owe so much to him, still govern the Society's affairs virtually unchanged, is a fine tribute to the farsighted vision of the founders among the foremost of whom must stand the statesman, Robert Moray, chief architect of the Society's incorporation and early life as a corporate body. No more devoted supporter of the Society could ever be elected to its presidential chair than this early occupant of it, nor has any single Fellow left a firmer imprint on the way The Royal Society has conducted its affairs for the past three centuries.[4]

Appropriately, Moray completes our roll call of the dozen founders of the Royal who were present at the 28 November 1660 meeting. But there are still a few from the Royalist camp worth mentioning because they played a significant part in founding the Society, even though they were not at Gresham College that day. The most important of these to the Royal was William Croone, who was thought of so highly by the group that they appointed him as 'Register' (that is, Registrar) of the Society at that foundation meeting, even though he was absent.

Croone was born in 1633, the son of a merchant living in London. He attended the Merchant Taylors' School, before going up to Emmanuel College, Cambridge, in 1647. After graduating in 1650, he became a Fellow of Emmanuel, where he stayed until 1659, when he became the Professor of Rhetoric at Gresham College, at the age of 26. In such a position at such a time he could hardly have failed to become one of the founders of the Royal Society. His duties as

4. Op. cit.

Registrar involved planning the programme for early meetings, and his first duty, according to Thomas Birch, was to find 'some discreet person skilled in shorthand writing' to take notes of the meetings. When the Society received its Royal Charter, the duties of the Registrar were subsumed in those of the first two Secretaries, Henry Oldenburg and John Wilkins; but Croone continued to play an active part in the Society.

Croone had wide-ranging scientific interests, and although he qualified as a physician in 1662 and had a successful medical practice (he acted as Censor of the College of Physicians in 1679, almost a hundred years after William Gilbert held the same post), the experiments he described to the Royal in the 1660s and beyond dealt with physical matters, such as the freezing of salt water and the compressibility of air, as well as subjects such as the circulation of the blood and the way muscles work. He was possibly the first person to use alcohol as a preserving fluid for organs and other animal tissues, and in 1662 he astounded the Fellows by showing how when the heart of a carp was cut out and divided into many pieces, each separate piece of heart muscle continued to contract and dilate for nearly a quarter of an hour. Samuel Pepys recounts the story of another dramatic demonstration in November 1662:

Dr Croone told me, that, at the meeting at Gresham College tonight, which, it seems, they now have every Wednesday again, there was a pretty experiment of the blood of one dogg let out, till he died, into the body of another on one side, while all his own run out on the other side. The first died upon the place, and the other very well, and likely to do well. This did give occasion to many pretty wishes, as of the blood of a Quaker to be let into an Archbishop, and such like: but, as Dr Croone says, may, if it takes, be of mighty use to man's health, for the amending of bad blood by borrowing from a better body.

Croone continued to play an active part in the meetings of the Society throughout his life, and in February 1684, just a few months before he died, he discovered the way in which water cooled to the freezing point expands slightly just before it turns to ice. He attended a meeting of the Council of the Society for the last time in July 1684, and died in October that year, soon after his 51st birthday. He spent almost half his life as one of the most regular attenders at the Society's meetings, and made real contributions both to the success of the Royal and to the development of science, although he has never quite received the recognition from history that his work deserves.

Sir Kenelm Digby never had a chance to make much of a mark on the Royal Society, because he was already 57 in 1660, and died in 1665. But his story is worth telling both because he had such an eventful life, and because he provides a link with the past. Digby was probably born on 11 June 1603, just three months after the death of Elizabeth I and four months before the death of William Gilbert. So a baby born before the death of Gilbert, the first scientist, lived to become a founding Fellow of the Royal Society. When you look at what he lived through, though, it seems almost miraculous that he lasted that long.

Digby came from a long line of soldiers and statesmen, although his grandfather, Everard Digby, was a philosopher and Fellow of St John's College in Cambridge. Unfortunately, Kenelm's father, Everard's son Sir Everard, reverted back to the type of his ancestors. He became a Catholic, fell in with a bunch of wild activists, and was executed in 1606, in his 28th year, for the part he played as one of Guy Fawkes' fellow conspirators in the Gunpowder Plot.

Kenelm's mother was just as fanatical a Catholic as her husband had been, but the boy was taken away from her and ended up in the care of his uncle, Sir John Digby, who was very much on the right side of the Catholic–Protestant divide, and a King's man through and through. In 1617, Sir John was appointed Ambassador to Spain, and took the 14-year-old Kenelm with him on a six-month visit to Madrid, where the boy learned the ways of Court. He also learned how to lip-read, studying with a priest who worked with deaf-mutes; wherever he went, Digby was always interested in learning new things. On his return to England in 1618, Kenelm went up to Gloucester Hall in Oxford, already, at only 15, a tall, handsome, and sophisticated young man, even by the precocious standards of the day. According to Aubrey, as an adult 'he was such a handsome person, gigantique and great voice, and had so gracefull Elocution and noble addresse, etc., that had he drop't out of the Clowdes in any part of the World, he would have made himself respected'.

In Oxford, Digby was strongly influenced by the mathematician and scientist Thomas Allen; as well as introducing him to mathematics, Allen showed him magnifying lenses, and told him about Galileo's astronomical observations with the telescope. But Digby was abruptly removed from Oxford in 1620, without taking a degree, because he had fallen in love with a beautiful girl, Venetia Stanley, of whom his mother disapproved, even though she was the daughter of a respectable gentleman, Sir Edward Stanley.

To give Kenelm a chance to forget Venetia, he was sent on a three-year Grand Tour of Europe. Unfortunately, in Paris he caught the eye of the dowager Queen, Marie de Médicis. Her husband Henri IV had been assassinated by a Catholic extremist in 1610,[5] and by 1620 Marie, now 47, was something of a spare wheel at Court, where her son, the 19-year-old Louis XIII, was King, but under the influence (to put it mildly) of Cardinal Richelieu.

Digby had the good sense not only to get out of Paris but to leave France entirely, leaving behind a story, to stop pursuit, that he had been killed in a brawl. Unfortunately, letters home informing his friends and relations that he was safe failed to arrive, and Venetia, instead of waiting for him, took the opposite course and became the mistress of the Earl of Dorset (and bore him children, if Aubrey is correct). Initially blissfully unaware of this, Digby settled in Italy, where he learned the language, studied, and became skilled at fencing. In 1623, he returned to his uncle in Madrid for another extended visit, and was there when the Duke of Buckingham made his unsuccessful attempt to arrange a marriage between Charles, the Prince of Wales, and Maria, the daughter of Philip III. Kenelm became involved in the negotiations as a go-between between Charles and Maria, partly because he was more their age than Sir John or Buckingham, and partly because it was thought that his Catholic family background might make him more acceptable to the Spanish. Even though the negotiations failed, when Digby returned to England in October 1623 he was knighted by James I as a reward for his efforts. He also became a Gentleman of the Privy Chamber to Prince Charles. Even better,

5. Although Henri had become a Catholic, he tolerated the Protestants, which the Catholic extremists objected to.

Venetia gave up her lover, married Sir Kenelm (who was now old enough to do so without his mother's consent), and produced two children (Kenelm in 1626 and John in 1627). But she died in 1633, probably of tuberculosis.

By then, Digby had gained fame throughout the country for a celebrated adventure at sea. In spite of the successes against the Spanish during Elizabeth's reign, the English navy was still far from being the dominant force that it would become in the eighteenth and nineteenth centuries, and in the mid-1620s English merchant shipping in the Mediterranean was suffering severely at the hands of pirates, mostly from France and Venice, but based in the Eastern Mediterranean with at least tacit support from Turkey. At this time, Charles I had recently come to the throne, but Digby had fallen out of favour with Buckingham (which certainly counts in his favour, given Buckingham's track record), and was therefore no longer as close to Charles, who was very much under Buckingham's influence. He decided to restore his position by sorting out the pirates, and fitted out two ships, the *Eagle* and the *Barque*, for the job.

This was taking a big chance. Digby's expedition is often referred to as privateering, but a privateer was a person (or ship) licensed by the King (or government) to attack enemy shipping. Digby had no such licence, and was technically a pirate himself; if the expedition had gone badly, he could have been disowned by the King. But it didn't. They set out from Deal in December 1627, took a few prizes from England's enemy The Netherlands en route, and attacked the pirates in the Bay of Scanderoon, just north of Cyprus, on 11 June 1628. The English triumphed in a fierce battle, filled their ships with booty, and promptly headed back to England.

According to legend, Digby amused himself on the voyage home by writing an essay on Spencer's *Faerie Queene*; he certainly did write such an essay (it was published in 1644), but whether he wrote it at that time is a moot point. What he certainly did write at that time was an account of the sea battle, which he had the foresight to send overland to England so that when the ships returned to Woolwich in February 1629 they were given a heroes' welcome. In truth, the battle wasn't in the same league as Francis Drake's famous singeing of the King of Spain's beard, but in the public eye at least it ranked somewhere near it. The King was delighted, and Digby's status at Court increased, while Buckingham's nose was put well out of joint.

If Venetia had lived, Digby would probably have settled down to a quiet life at home. But in 1634, the year after her death, he gave his valuable collection of books and papers to the Bodleian Library in Oxford, and spent the next four years on the continent, mostly in Paris. He became friendly with René Descartes, reviving his own interest in mathematics, and collected the basis of another library of books. When he came back to England, he became part of the circle of friends of the Queen, Henrietta Maria. As a Catholic and sister of Louis XIII, the Queen was rightly suspected of plotting to restore Catholicism in England. Digby got involved in these plots, and came under the suspicion of Parliament in the run-up to what became the Civil War. This time, seeing discretion as the better part of valour, he went back to France. There, he challenged a French nobleman to a duel after the Frenchman publicly insulted Charles I, and duly ran him through the heart, having lost none of the fencing skills he had learned in Italy. The King of France could hardly do anything except pardon Digby for this act in defence of his brother-in-law the King of England, but Digby had to leave Paris for the safety of Flanders (with an escort of 200 of the King's soldiers), and then on to England, where he was promptly imprisoned, in 1641.

Digby was still in prison at the time of the Battle of Edgehill and the opening phase of the Civil War, but was released in July 1643, on condition that he left England. He went back to Paris, where the heat had died down, the same year that the 5-year-old Louis XIV (later to be known as the 'Sun King') came to the throne. Cardinal Mazarin was the power behind the throne, and English Catholics were welcome, even if, as in Digby's case, they had a somewhat chequered past. The first thing Digby did after settling in Paris was to arrange for the publication of a huge encyclopedic work, known as *Two Treatises*, much of which must have been prepared during his years in prison. This dealt with the body and the soul (hence the title), and the part on the body dealt in particular with embryology, the nervous system, and the blood vessels (the vascular system). The book provided the first full account in English of Harvey's ideas on the circulation of the blood, and it is clear from the book that Digby knew Harvey well and had discussed these ideas with him. Although not original, this was a major contribution to what would now be called the public understanding of science, and editions of the book were published in London in 1645, 1658, 1665, and 1669.

Science had to take a back seat, however, while Digby became embroiled in various Catholic plots over the next decade. After Henrietta Maria fled from England to France in 1644, she appointed Digby as her ambassador to the Vatican, where he persuaded the Pope to part with 20,000 crowns to help outfit a Catholic army (headed by Digby) to fight against the Puritans in England; but the army never materialized, and nothing came of the other schemes either. After Charles I was beheaded in 1649, Digby thought it would be safe to return to England, but he was promptly banished once again, by the ruling Council of State, and only finally allowed to return home at the end of 1653, when he was 50, having made promises that he would not cause trouble. He got to know Cromwell, and even tried (unsuccessfully) to persuade him to soften his policy on Catholics.

Digby made his last visit to the continent in 1657, taking a holiday in the spa town of Montpelier to seek a cure for the kidney stone that afflicted him. While he was there, he gave a lecture at the Montpelier Academy of Sciences on the value of the powder ferrous sulphate (now known to have antiseptic properties) in medicine; this was later published in French, and then in English under the title *Powder of Sympathy*. It was after Digby returned from this trip to France in 1658 that he became friendly with John Wallis, Robert Boyle, and Robert Hooke, and wrote a series of letters to Wallis in which he set out his ideas for establishing a scientific society where the latest discoveries could be discussed. He was very much one of the group of Royalists who welcomed the return of the King, and whose presence among the group of experimental philosophers encouraged the King to take them seriously. Although he was not present at the meeting at Gresham College on 28 November 1660, Digby had attended earlier meetings of the group, and on 23 January the following year, just eight weeks after the inaugural meeting, he read to the fledgling Society his paper *A Discourse Concerning the Vegetation of Plants*. This was the first publication formally authorized by the new (and as yet unnamed)[6] Society. It was also Digby's last publication. He served on a committee set up by the Royal to look into the application of science (we would now say technology) to the navy, and his interest in plants was invaluable when the Society became involved in efforts to promote the growth of timber to replenish stocks lost to ship building. But his health soon began to deteriorate, and he died in 1665, on 11 June, which was said to be his birthday.

The calendrical coincidence prompted the following epitaph, from Richard Farrar:

> Under this Tomb the Matchless *Digby* lies;
> *Digby* the Great, the Valiant, and the Wise;
> This Ages Wonder for His Noble Parts;
> Skill'd in Six Tongues, and Learn'd in All the Arts.
>
> Born on the Day He Dy'd, Th'Eleventh of June.
> And that Day Bravely Fought at *Scanderoon*.
> 'Tis rare, that one and the same Day should be
> His Day of Birth, of Death, and Victory.

As well as being his last, *Vegetation of Plants* was also Digby's only really original contribution to science, and is arguably the first scientific publication on horticulture. He looked at the question of what a plant needs in order to live, and argued that saltpetre (nitrous salt) is an essential ingredient in the soil, along with water and air. He drew an analogy between the sprouting of a seed and the development of a chick embryo. He said that 'there is in the Aire a hidden food of life' and that 'languishing' plants in a sealed glass container could be revived by providing them with good air. 'My own eyes are witnesses of the wonderfull corporifying of it,' he said, emphasizing from the very first the importance to the Royal Society and its Fellows of seeing things for themselves, and not relying on hearsay or tradition as evidence – the reason for their eventual choice of motto, *Nullius in Verba*, which translates literally as 'nothing in words' but should be taken as meaning 'take no man's word for it'. Sir Kenelm Digby may

6. Not *quite* unnamed; on the title page of the published version of the lecture, it is described as having been presented 'At a Meeting of the Society for promoting Philosophical Knowledge by Experiments'.

not have been a typical founder member of the Royal Society, but his attitude towards science certainly typified that of the Society.

The last of the early members of the Royal Society we want to mention, John Evelyn, made only modest contributions to science, but is worthy of inclusion both because of his famous diary, which gives us insight into many of the activities of the Society and its Fellows in the early days, and because he was the first person to use the name 'Royal Society' in print, and may have been the person who came up with the name.

Evelyn came from another of those wealthy families that had risen to success during the reign of Elizabeth I. His grandfather sowed the seeds of the family fortune when he introduced the manufacture of gunpowder into England, and his father rose to become Sheriff of the combined counties of Surrey and Sussex in 1633–4 – as we have seen, a not unusual story of upward mobility in those days. John Evelyn was born in 1620 at the village of Wotton, near Dorking, in Surrey; but he was largely brought up by his mother's family at Lewes, in Sussex. He went to school in Lewes, then at the age of 17 to Balliol College, Oxford. Although he was also admitted to the Middle Temple, this was not to study law but to provide social grooming and useful contacts for his later life. Evelyn later said that neither school nor university had provided him with much of an education, although he was an observant and inquisitive individual who developed a keen interest in the visual arts.

Evelyn's father died in 1640, and the family estate at Wotton passed into the care of John's elder brother. When the Civil War broke out, the estate was in an area controlled by Parliament. If either brother had shown active support for the Royalist cause, the estates would have been seized, so both brothers carefully avoided getting involved. In November 1643, John Evelyn decided to get clear of any possible trouble, and travelled through France to Italy, where he studied anatomy and physiology in Padua at the end of 1645 and the early months of 1646. On the way back towards England, he briefly studied chemistry and horticulture in Paris, where, in 1647, he also got married, to Mary Browne, the daughter of Sir Richard Browne, Charles I's diplomatic agent in France.

As the situation in England stabilized, Evelyn and his wife spent part of their time in England and part in France, before settling in 1652 on Sir Richard's estate at Deptford, which was already an important naval dockyard in those days (Samuel Pepys often visited it during his time with the Navy Office), although still a village outside the sprawl of London. Over the next seven years, Evelyn lived quietly and avoided getting involved in any Royalist plots; his wife produced five babies, all boys, during the 1650s, but only two of them survived; three daughters were born later. Evelyn kept busy by improving his father-in-law's estate (most notably by planting trees), translating the first book of Lucretius, and taking a broad interest in science. He visited Oxford in 1654 and 1656, meeting Wilkins, Wren, Boyle, and Neile, among others. He was particularly impressed by Boyle.

With the Restoration, Evelyn was assured of a welcome at Court, both because of his known Royalist sympathies and because of the position of his father-in-law under Charles I. He served the new regime on several commissions involved in such diverse projects as improving the streets of London, overseeing the Royal Mint, and looking after the sick and wounded from the two Dutch Wars, 1664–7 and 1672–4.[7] Alongside this, Evelyn, who had been a regular attender at the Gresham College meetings even before 28 November 1660, took an active interest in the Royal Society for the next quarter of a century. He served as a Member of Council fifteen times and did a spell as Secretary in 1672–3. He eventually inherited the estate at Wotton when his brother died in 1694, and lived on until 1706, when he died at the age of 85.

Evelyn's most visible contribution as a Fellow of the Royal Society resulted from a request by the Commissioners of the Navy, in 1662, for a study of the forestry problem, which had been exacerbated by the great expansion of the navy during the Interregnum. Evelyn took on the task of putting the thoughts of the committee set up by the Royal in order and seeing them into print; the result, published in 1664, was a book on the propagation of timber, which is usually known by the short version of its title, *Sylva*. The book appeared in

7. These wars are generally remembered as a futile waste of men and resources; but it was among the terms of the peace of 1667 that the Dutch ceded New Amsterdam, soon renamed New York, to England.

successively larger revised editions over the next four decades, the second in 1670, the third in 1679, and a fourth in 1706, just after Evelyn died. A fifth edition came out in 1729, and in 1776, more than a hundred years after the original edition, an annotated version of the fifth edition appeared; this in turn went into four further printings, the last in 1825, just as the wooden ships of the Royal Navy were about to be replaced by ironclads and vessels with steel hulls. What makes this all the more remarkable is that the original book is *not* remarkable, presenting some straightforward advice about the cultivation of trees along with a mish-mash of anecdotal material lifted from other sources. But there is no denying its success.

The first published reference to the Royal Society by that name came in 1661, when Evelyn translated a book called *Instructions Concerning Erecting of a Library*, by Gabriel Naudé, from the French. In his dedication of the book, to the Lord Chancellor, Lord Clarendon, Evelyn sings the praises of the Society, and twice refers to it by its now familiar name. Since the Society was consciously set up in the image of the Foundation imagined by a previous Lord Chancellor, this was a particularly nice touch. But now we have met all the men who played key roles in getting the Royal Society started, it's time to see how they did it, and what they did with it.

# 6

# The Man Who Made It Work

The founders of the Royal Society were mostly a remarkable group of men who had the enthusiasm and the connections that made it possible for the Society to get its start. But they all had other interests and careers to pursue alongside their interest in the fledgling society. The only founding Fellow who could be regarded as being anything like a full-time scientist was Robert Boyle – and even he had his duties and obligations as a member of the aristocracy to carry out, quite apart from the fact that in the early 1660s he was still based in Oxford. The early enthusiasm which led to the founding of the Society might well have faded away, and the Society with it, if it had not been for the indefatigable work of one man, Robert Hooke, who was recruited two years after the 28 November 1660 meeting, but did more over the next two decades than any other single person to make the Royal a truly scientific society. In the words of J. G. Crowther, 'without Hooke the Society might have disintegrated'.

To see why, and to put Hooke's contribution into perspective, we need to pick up the story of the Society itself, now that we have met its founders, after that inaugural meeting. The primary source of information is Thomas Sprat's *History*, published in 1667. Sprat, as we have mentioned, was another of Wilkins' protégés – he had gone up to Oxford in 1651, entering Wadham College two years after Christopher Wren; at the time he wrote his book he was a Fellow of Wadham, and although the *History* was his only significant contribution to the Royal, he went on to become a chaplain first to the Duke of Buckingham, then to Charles II, before ending up as Bishop of Rochester. Because of previous misunderstandings, it's worth emphasizing just how official, and authoritative, Sprat's *History* was.

(1) *Numb.* 1.

# PHILOSOPHICAL
## *TRANSACTIONS.*

---

*Munday, March* 6. 166$\frac{4}{5}$.

---

The Contents.

*An Introduction to this Tract. An Accompt of the Improvement of* Optick Glaſſes *at* Rome. *Of the Obſervation made in* England, *of a Spot in one of the Belts of the Planet* Jupiter. *Of the motion of the late* Comet *prædicted. The Heads of many New Obſervations and Experiments, in order to an Experimental* Hiſtory of Cold; *together with ſome* Thermometrical *Diſcourſes and Experiments. A Relation of a very odd Monſtrous* Calf. *Of a peculiar* Lead-Ore *in* Germany, *very uſeful for Eſſays. Of an* Hungarian Bolus, *of the ſame effect with the* Bolus Armenus. *Of the New* American *Whale-fiſhing about the* Bermudas. *A Narative concerning the ſucceſs of the* Pendulum-watches *at Sea for the* Longitudes; *and the Grant of a* Patent *thereupon. A Catalogue of the Philoſophical Books publiſht by* Monſieur de Fermat, *Counſellour at* Tholouſe, *lately dead.*

*The Introduction.*

WHereas there is nothing more neceſſary for promoting the improvement of Philoſophical Matters, than the communicating to ſuch, as apply their Studies and Endeavours that way, ſuch things as are diſcovered or put in practiſe by others; it is therefore thought fit to employ the *Preſs*, as the moſt proper way to gratifie thoſe, whoſe engagement in ſuch Studies, and delight in the advancement of Learning and profitable Diſcoveries, doth entitle them to the knowledge of what this Kingdom, or other parts of the World, do, from time to time, afford, as well

A of

*The first page of the first number of the* Philosophical Transactions *of the Royal Society*

It's hard to understand why any misunderstanding on this point could have arisen, because he makes his position clear himself. Describing the source of his information and his role in seeing it into print, he says of the Royal Society that

They had at first a *Register*, who was to take Notes of all that pass'd; which were afterwards to be reduc'd into their *Journals* and *Register Books*. This Task was first *perform'd* by Dr *Croone*. But they since thought it more necessary, to appoint two *Secretaries*, who are to reply to all Addresses from abroad, and at home; and to publish whatever shall be agreed upon by the *Society*. These are at present, Dr *Wilkins*, and Mr *Oldenburg*, from whom I have not usurp'd this first imployment of that kind; for it is onely my hand that goes, the substance and direction came from one of them.

'It is onely my hand that goes.' In other words, Sprat is responsible for physically putting the words on paper; but what he writes is what Wilkins or Oldenburg tell him to write, in their official capacity as Secretaries of the Royal. As Margery Purver sums the situation up, 'Sprat was not speaking for himself nor for any other private person, but for the Royal Society as an institution, which considered this book to be its special concern, the first comprehensive and public account of its origin, policy and business.'

So we can confidently accept that Sprat's description of the philosophy behind the Royal is the official one:

Whatever they have resolv'd upon; they have not reported as *unalterable Demonstrations*, but as *present appearances*: delivering down to future Ages, with the good success of the Experiment, the *manner* of their progress, the *Instruments*, and the several differences of the *matter*, which they have apply'd: so that, with their mistake, they give them [future generations] also the means of finding it out. To this I shall add, that they have never affirm'd any thing, concerning the cause, till the trial was past: whereas, to do it before, is a most venomous thing in the making of *Sciences*: for whoever has fix'd on his *Cause*, before he has experimented; can hardly avoid fitting his *Experiment*, and his Observations, to his own *Cause*, which he had before imagin'd; rather than the *Cause* to the truth of the Experiment it self.

There have been many scientists since the 1660s who could have learned from that example, but there has seldom been a clearer statement of the need to fit hypotheses to the facts, rather than distorting the facts to fit preconceptions. As the passage also makes clear, the founders of the Royal were well aware that they were only at the beginning of the task spelled out by Bacon, and they also followed the Baconian precept of gathering as much data as possible, about everything they could, for the benefit of future generations:

The *Society* has reduc'd its principal observations, into one *common-stock*; and laid them up in publique *Registers*, to be nakedly transmitted to the next Generation of Men; and so from them to their Successors. And as their purpose was, to heap up a mixt Mass of *Experiments*, without digesting them into any perfect model; so to this end, they confin'd themselves to no order of subjects; and whatever they have recorded, they have done it, not to compleat Schemes of opinions, but as bare, unfinish'd Histories.

Sprat is quite explicit about the origins of the Royal (and, remember, he is merely writing what Wilkins and Oldenburg tell him to write):

It was . . . some space after the end of the Civil Wars at *Oxford*, in *Dr Wilkins* his Lodgings, in *Wadham College*, which was then the place of Resort for Vertuous, and Learned Men, that the first meetings were made, which laid the foundation of all this that follow'd . . . The principal, and most constant of them, were Doctor *Seth Ward*, the present Lord Bishop of *Exeter*, Mr *Boyl*, Dr *Wilkins*, Sir *William Petty*, Mr *Matthew Wren*, Dr *Wallis*, Dr *Goddard*, Dr *Willis*, Dr *Bathurst*, Dr *Christopher Wren*, Mr *Rook*: besides several others, who joyn'd themselves to them, upon occasions.

Sprat then goes on to explain how the various members of the group ended up in London, where they were joined for their meetings at Gresham College by

The Lord *Viscount Brouncker*, the now Lord *Brereton*, Sir *Paul Neil*, Mr *John Evelyn*, Mr *Henshaw*, Mr *Slingsby*, Dr *Timothy Clark*, Dr *Ent*, Mr *Ball*, Mr *Hill*, Dr *Crone*: and divers other Gentlemen, whose inclinations lay the same way.

With the influx of what we have called the King's Men, as soon as the group decided to form themselves into a society at the end of 1660, they were in a position to obtain the kind of formal recognition of their status that Bacon had referred to in his *New Atlantis* as a

'King's Charter' – recognition which would confer (perhaps) 'gift of revenue, and many privileges, exemptions and points of honour'. Of course, the newly restored Charles II was broke, and there was no hope of financial support from him; but a Royal Charter would not only enhance the prestige of the Society, but would give it a legal status offering financial advantages (such as freedom from some forms of taxation), the right to employ staff and to buy property, and the authority to publish books and other works under its own imprimatur. It might also indirectly provide income, since membership of a Royal Society might be attractive to rich gentlemen with a passing interest in science, whereas they would not be eager to pay annual fees to be members of a mere club or debating society, however scientific. It is one of the most remarkable features of the success of the Royal that it achieved this with no official financial support, and did indeed fund its activities from the revenues provided by its Fellows – albeit not without considerable difficulty, as we shall see.

The Society continued to meet at Gresham College after 28 November 1660, largely because there was nowhere else they could meet. An approach to the College of Physicians to provide them with accommodation failed, but by the end of the following March they had at least obtained the use of a room at Gresham, instead of having to meet in the private rooms of one of the Gresham professors. By then, Sir Robert Moray had been installed as President (on 6 March 1661). We have seen how he had been able to report back to the Society at only its second meeting, on 5 December 1660, that the King approved of their activities and wished to encourage them; now, as President and with the King's ear, he was to be the key player in getting the first Royal Charter for the Society, in 1662. Along with the various duties and privileges which the Charter gave the Society (including the right to bodies of executed criminals for dissection), the letters patent gave it the name the Royal Society, and named Brouncker as the first President under the new Charter.

For some reason, though, the Council of the newly Royal Society decided that they weren't happy with the Charter. They immediately requested changes, as a result of which a second Charter was issued in 1663. The principal differences were that in this Charter the name of the Society was given as 'the Royal Society of London for Promoting

Natural Knowledge', the King was formally acknowledged to be both founder and patron of the Society, and a coat of arms (with the motto *Nullius in Verba*) was granted. In effect, the first Charter never came into force, and the establishment of the Royal Society in its final form legally dates from the second Charter, which received the Royal Seal on 23 April 1663, and was presented to the Society on 13 May 1663. This was therefore the first meeting of what we know as the Royal Society. A few days later, in a letter to the Dutch scientist Christiaan Huygens, Moray wrote:

> We begin now to work on establishing our Society with more energy than hitherto, because the King's patent was sent to us five or six days ago, which establishes it as a Corporation with various privileges. Such a charter is necessary, according to the laws of this country, in order for it to be able to receive donations, to give it legal status etc. So that we are now concentrating our efforts on other necessary means for the prosecution of the schemes we proposed for ourselves, as the constitution of the Society, which will have an impact on funds needed to pay the cost of the experiments.[1]

Clearly *a* charter was essential, although it's hard for modern eyes to see why the founders should have cared so much about the specific points mentioned above that they requested a second charter from the King. But the reference to 'promoting natural knowledge' and the motto *Nullius in Verba* both emphasize the scientific nature of the Society, and the stronger emphasis on the Royal connection was clearly desirable. There may also have been different factions at work among the Fellows which we now know nothing of, with different ideas about the exact wording of the Charter. But the politics need not concern us. What matters much more, in terms of the revolution in science, is that while the political manoeuvring to obtain a Royal Charter was going on, the Society had first of all got itself into a bit of a mess with its experimental activities, and then found just the man to get them out of that mess.

Because the Society was utterly dependent on its income from membership fees, from the outset it had to attract wealthy men to its

1. Translation from the original French after Lisa Jardine, *The Curious Life of Robert Hooke*.

ranks. Some idea of just how wealthy they had to be is given by the size of the joining fee, which soon increased from the original 10 shillings to £2, and the annual subscription of £2 12s, or a shilling a week. The rich amateur scientists who joined the Society wanted to be entertained by the new science, not just to be part of an intellectual exercise. Nevertheless, they were an important part of the intellectual exercise, acting as trustworthy observers of the experiments they witnessed. As Sprat emphasizes, there was an important role for 'plain, diligent and laborious observers: such, who, though they bring not much knowledge, yet bring their hands, and their eyes uncorrupted . . . and can honestly assist in the *examining* and *Registering* what the others represent to their view'. *Nullius in Verba*, in fact.

The trouble was, who could carry out the experiments which would both entertain the more dilettante Fellows and provide valuable scientific insights for them to witness? At first, the active experimenters who formed the core of the Society were expected to take turns at the task. At the second meeting, on 5 December 1660, rules were laid down to cover the presentation of these demonstration experiments, and the name 'curator' was chosen for the person performing the demonstration. But this simply did not work. There were not enough people able and willing to offer new work to fill up the weekly slots, and there was no continuity from one week to the next.

Hooke attended meetings of the Society in 1661 and 1662 as Boyle's assistant (they stayed at Lady Ranelagh's house in Chelsea while visiting London), and demonstrated some of their joint experiments, but his name scarcely appears in the official or unofficial records of the time. Evelyn, for example, refers to a meeting on 25 April 1661 where 'divers experiments in Mr Boyle's Pneumatic Engine' were demonstrated, without mentioning that it was in fact Mr Hooke who did the demonstrations (and, indeed, had built the air pump!). The pump used in these demonstrations was actually the original Boyle–Hooke pump from Oxford, which Boyle had given to the Society since Hooke had made him an improved model; but nobody at the Society had the skill to make it work properly, and it largely languished until Hooke eventually became the Curator of Experiments.

Semi-anonymous or not, Hooke's position as Boyle's right-hand man drew him to the attention of the 'virtuosi' who formed the core of the Society (many of whom, including Wilkins, already knew him from their Oxford days), and it became clear that they needed somebody with just his skills to work for them full time. On 12 November 1662, Moray formally proposed that Hooke be appointed as Curator of Experiments, 'to furnish them every day, on which they met, with three or four considerable experiments'. At the next meeting the proposal was accepted unanimously, with Hooke being told that he should 'come and sit among them, and both bring in every day [that is, every day the Society met] three or four of his own experiments, and take care of such others as should be recommended to him by the Society'. As if that were not enough (considering that the meetings were held every week), he would also be expected to carry out other tasks as the Fellows required. The agreement of the Society to the arrangement was a formality, since Moray had already obtained both Boyle's and Hooke's agreement. This act of generosity on Boyle's part extended to providing continued financial support for Hooke while the Society tried to get its affairs in order – Hooke continued to use Lady Ranelagh's house as his London base for a time – which involved some creative financial accounting. The snag was that this meant that in addition to the demands placed on him by the Society, Hooke, now 27, continued to have obligations to Boyle (by now his friend as well as his employer) which made further demands on his time.

Clearly, the situation would have to be regularized as soon as possible. Although a salary was implied, at first the Society simply had no means to pay Hooke, although he took up his duties immediately, and brought about a marked improvement in both the quality and quantity of the experiments being demonstrated at the weekly meetings. His first reward from the Society was not financial; in June 1663, Hooke was elected as a Fellow, with the usual subscription fees waived. The same year, as we have mentioned, he was admitted as MA in Oxford, in spite of never completing his BA. All this marked his acceptance as one of the scientific elite, a member of the virtuosi that made up the inner circle of the Society. But it did not release him from his obligations to the other Fellows, and alongside a great outpouring of scientific work in the early 1660s, of which more shortly, Hooke was constantly looking for a regular, independent income.

An ideal opening appeared in May 1664, when Isaac Barrow resigned the Gresham Chair of Geometry. Hooke applied for the post, but it went to one Arthur Dacres, a physician at St Bartholomew's Hospital, in spite of Hooke's application being supported by the Royal. Shortly after this disappointment, Hooke was gloomily contemplating the implications over a drink in a pub with John Graunt, another Fellow of the Royal Society, when they were joined by Sir John Cutler, another of Graunt's friends. Cutler, originally a member of the Grocers' Company, had made a fortune in banking, and was so touched by Hooke's story that he offered to pay £50 a year to Hooke for life (the same as the salary of a Gresham professor) in return for lectures at Gresham College on the history of trades (including the practical application of science to trades). Unfortunately for Hooke, when Cutler made the offer formally it came through the Royal Society, and the Royal immediately saw how to make use of his generosity for their own ends. On 27 July 1664, a Council meeting of the Royal decided

That at the first opportunity Mr Hooke be put to the scrutiny for the place of curator. That he should receive eighty pounds per annum, as curator to the society by subscriptions of particular members, or otherwise: That he forthwith provide himself of a lodgings in or near Gresham-College: And That these orders and votes be kept secret, till Sir John Cutler shall have established Mr Hooke as professor of the histories of trades.

So when Cutler's money did start coming in, all the Royal had to do was add £30 a year to top up Hooke's salary! By these machinations, the Royal ensured that more than half of Hooke's salary as Curator came from Cutler, so he had extra work to do but no extra money. Worse, Cutler proved somewhat erratic in making the payments, and in the end Hooke had to resort to legal action (with no help from the Royal) to get what was due to him, adding yet another time-consuming burden to his load.

But it wasn't all gloom. Hooke moved into rooms at Gresham College in September 1664, even though he wasn't yet a Gresham Professor. There was, though, by then every hope that he soon would be. In the summer of 1664 the Royal had learned that the election of Dacres to the Chair of Geometry had been rigged. The committee that made the appointment had nine members, with the Lord Mayor of London, Sir Anthony Bateman, sitting in as an observer. The committee had voted 5:4 in Hooke's favour, but Bateman had then voted himself (which he was not entitled to do), making a tie, and then said that he had the casting vote in such a situation, and declared Dacres had won. Since there were two other members of the Bateman family on the committee, and they had both voted for Dacres, it became clear they had been involved in a fix.

On this occasion, the Royal was able and willing to help Hooke make his case, to the new Lord Mayor, Sir John Lawrence. The upshot was that in March 1665, at the age of 29, Hooke did become Gresham Professor of Geometry, with comfortable rooms in the college that included a library, parlour, and two smaller rooms on the ground floor, with a garret above for a servant and three cellar workshops below. He was now financially secure (indeed, very well off, if all the money due to him was paid on time) and settled in his various roles as Gresham Professor, Cutlerian Lecturer, and Curator of Experiments to the Royal Society – any one of which would have been enough for a normal man to undertake. Christopher Wren, himself a high achiever, wrote to Hooke in April 1665 that 'I know you are full of employment for the Society wch you allmost wholly preserve together by your own constant paines'. Even if you have reservations about regarding Robert Boyle's paid assistant as the first professional scientist, there can be no doubt that in March 1665, completely free of the need for Boyle's patronage, Hooke achieved that status.

By that time, Hooke's most famous publication, *Micrographia*, had been in print for two months. This was one of the first truly popular accounts of science (described by Pepys, who sat up until 2 a.m. reading one of the first copies, as 'the most ingenious book that ever I read in my life'), and was by no means restricted to reporting what Hooke had seen with the aid of his microscopes, although that was the main theme. To a large extent, the book summed up Hooke's scientific work since the late 1650s, some of it carried out in collaboration with Boyle, but a lot of it entirely his own. There was, though, one significant omission. Like many of his contemporaries, Hooke was fired by the idea of finding a method for determining longitude at sea, and getting rich from the invention. This was the key problem in navigation, and would be of immense value to any seafaring nation; his attacks on the problem coloured much of Hooke's scientific work, but he was secretive about many aspects of this, for fear of having his ideas stolen, and this meant that in the end he did not always receive proper credit for this aspect of his work, which came close to claiming the ultimate prize.

The key to determining longitude at sea was to find an accurate means of telling the time, according to some established standard, so that you could compare the time of local noon (or midnight) with your standard time, and thereby determine how far east or west of the home port you were. One way to do this, in principle, was by accurate astronomical observations, for example, of the moons of Jupiter. The moons follow regular, predictable orbits, so it is possible to predict what pattern the moons will be in, relative to one another, at a certain time according to clocks in, say, London. If it were possible to make accurate astronomical observations of this kind from a ship at sea, that would give you a way of telling the time in London even if you were thousands of miles away. The idea was fine in theory but horrendously difficult in practice, although that didn't stop several astronomers working on it. It was largely because of this practical need for a way to determine longitude at sea, not simply due to an abstract interest in science, that Hooke spent a lot of his time developing accurate astronomical instruments, known as sextants and quadrants. The work proved a boon to astronomers, but no use to mariners.

The other obvious way to determine longitude was to carry on board ship a clock (or clocks) set to the time in the home port, and which kept accurate time throughout the voyage. Then, all you would have to do would be to compare the time on the clock with the time of local noon, when the Sun is highest in the sky, to work out your longitude. The first reasonably accurate pendulum clock was patented by Christiaan Huygens in 1657; but there are obvious difficulties with getting a pendulum clock to work accurately on a ship rolling about at sea. Hooke (harking back to his boyhood skill at making the wooden clock) developed a mechanism driven by the regular oscilla-tion of a spring, rather than pendulum, which could be used to make an accurate clock or watch. This use of a vibrating spring to provide the regular ticking of the watch is, of course, quite different from the use of a wound-up spring to provide the power which keeps the watch ticking. Hooke was so convinced that he had solved the longitude problem that some time late in 1663 or in 1664 (the exact date is not recorded) Robert Boyle arranged for a meeting between himself, Hooke, Moray, and Brouncker to agree the terms under which Hooke would divulge his secret. But the others insisted that although Hooke would be rewarded handsomely and granted a 14-year patent if the watch worked, anyone who improved on his design should receive the income from the improved watch. Hooke refused to accept this, saying that once the basic idea was out it would be easy for someone to make minor improvements and claim the income. In the end Hooke's design for the marine watch was never made public, and the secret died with him, although he did make several watches, one of which he presented to the King.

In the 1670s, all this led to a bitter row between Hooke and Huygens, who also invented a spring-driven watch, over priority (see page 219). For what it is worth, it is now clear that Hooke's original invention, probably in 1660 or 1661, was certainly the first spring-driven watch, but Huygens' invention was made entirely independently. But Hooke's marine watch probably would not have solved the longitude problem. Modern analysis of Hooke's surviving papers does not reveal exactly what his secret improvements to the basic design were, but the evidence suggests that although the watch may have kept good time in spite of the motion of a ship, he had forgotten to take account of the problems that would be caused by changes in temperature affecting the behaviour of the spring on the ship's travels. This was the key improvement that John Harrison made, in the 1720s.

Hooke's unsung invention of the spring-driven watch is another example of how his scientific investigation of the world, like that of many other early Fellows of the Royal Society, was driven by practicality. The one reason why schoolchildren still come across Hooke's name is because the law of elasticity is named after him. This comes across as a rather mundane piece of useless information – that the amount by which an elastic object stretches is proportional to the force pulling on it. But Hooke discovered this simple law directly through his work on springs in connection with his attempt to find a way to make a chronometer that would keep accurate time at sea and make him rich.

There is no space here to provide more than the briefest account of Hooke's scientific activities in the first half of the 1660s. Many of his demonstrations centred around the air pump, which he induced to work as well as ever. He also demonstrated to the Fellows, in December 1662, the way in which a hollow glass sphere would float on top of a container of cold water but would sink when the water was heated. He realized that warmer water is less dense than cold water and that by heating a tank of liquid from below you could make what are now called convection currents which would stir up the liquid and distribute the heat throughout the tank. But a proper understanding of convection only came after it was rediscovered by Benjamin Thompson (Count Rumford) in the early 1800s.

Hooke was involved in developing a kind of inverted bucket with a hinged double lid to obtain samples of sea water from different depths, and in several attempts to find ways of providing divers with air, the most successful of which involved providing the diver, on the sea bottom, with air sealed in lead boxes. In the autumn of 1663, at the behest of the Royal Society, he began to keep daily records of the weather. As with all his activities, Hooke threw himself wholeheartedly into the task, setting out systematically to record the wind speed and direction, humidity, temperature, air pressure, and the appearance of the sky. He was always interested in developing a full understanding of anything he investigated, and was not one to take a passing interest in a subject and then drop it. He invented a portable mercury barometer, and then the first wheel barometer, where the rise and fall of pressure is measured by the movement of a pointer around a dial, like a clock face. The interest in weather lasted for the rest of his life, making Hooke arguably the first meteorologist, but only a year after receiving his brief from the Royal he would write to Boyle that:

I have . . . constantly observed the baroscopical index . . . and have found it most certainly to predict rainy and cloudy weather, when it falls very low; and dry and clear weather, when it riseth very high, which if it continues to do, as I have hitherto observed it, I hope it will help us one step towards the raising a theoretical pillar, or pyramid, from the top of which, when raised and ascended, we may be able to see the mutations of the weather at some distance before they approach us, and thereby being able to predict, and forewarn, many dangers may be prevented, and the good of mankind very much promoted.

But that dream would not begin to come true until the work of Robert FitzRoy, the founder of the British Meteorological Office, in the second half of the nineteenth century.

Through his interest in the nature of air and in respiration, Hooke came close to making one of the most important discoveries in chemistry. In spite of Harvey's work, in the 1660s nobody had much idea *why* it was necessary for blood to travel between the heart and the lungs, and what happened when air and blood mixed. There was even doubt about whether the pumping action of the lungs was simply a means of sucking in air, or whether it was necessary to force the blood and air to mix. In a rather gruesome vivisection experiment involving a dog, in November 1664 Hooke demonstrated how the animal could be kept alive by pumping air into its lungs from a bellows; but he wrote to Boyle that 'I shall hardly be induced to make any further trials of this kind, because of the torture of the creature'.

Hooke had earlier compared combustion and respiration by investigating what happened to a lighted candle and a chick placed in a sealed vessel. People knew that something in the air was necessary to sustain life, and also to sustain combustion. Hooke now made the intellectual leap of arguing that gunpowder, which burns without air, must contain the same substance that is in air that sustains combustion. Early in 1665, he showed (to the Fellows of the Royal) that neither charcoal nor sulphur, two of the three ingredients of gunpowder, would burn without air, but that they would if mixed with the third ingredient, saltpetre (which we now call potassium nitrate). In *Micrographia* Hooke wrote that combustion in air is 'made by a substance inherent, and mixt with the Air, that is like, if not the very same, with that which is fixt in Salt-peter'.

We now know that it is indeed always the very same substance involved in combustion – oxygen from the air, or, in the case of gunpowder, oxygen released by the potassium nitrate. But this modern view only emerged from the work of Joseph Priestley and Antoine Lavoisier in the last quarter of the eighteenth century, more than a hundred years after Hooke's work with gunpowder. Until then, in spite of his clear insight, progress would be hampered by the widespread acceptance of the idea that combustion involved the *release* of a substance called phlogiston from a burning object, rather than something from the air combining with the material of the burning object.

In December 1664 (too late for the observations to appear in *Micrographia*), Hooke, in London, and Wren, in Oxford, had been asked by the Royal to make observations of a new comet. The two astronomers came to the tentative conclusion that the comets did not move in straight lines, as had been widely supposed, but followed curved paths, as the unsung English clergyman and astronomer Jeremiah Horrocks had suggested some thirty years earlier. Pepys records that on 1 March 1665 he heard Hooke give a lecture at Gresham College 'about the late Comett, among other things proving very probably that this is the very same Comett that appeared before in the year 1618, and that in such a time [46 years] probably it will appear again – which is a very new opinion'. A second comet appeared in March 1665, and further observations strengthened the idea of curved cometary orbits. The discovery implied that comets were moving around the Sun in the same sort of way that planets were. Although this did not immediately lead to the conclusion that all these objects are under the influence of a force emanating from the Sun, Hooke seems to have been already thinking about the nature of gravity and the possibility that it extended to planets and comets as well as to objects on Earth.

Hooke had carried out many experiments with falling objects, extending Galileo's studies of acceleration, and tried to find out if things weighed less at higher altitude by making measurements from the top of Westminster Abbey. He used the most accurate balance scales available, with two identical weights, one on one pan of the scales and the second attached to the other pan by a very long thread, so that the weight itself was at ground level. But he found no change in weight at different heights, even when the experiment was repeated from the taller tower of St Paul's Cathedral. He carried out experiments with powerful air guns, estimated the height of the atmosphere from barometric measurements (improving on an earlier estimate by Boyle), and was the first person to explain the twinkling of starlight as due to the refraction of light in the atmosphere of the Earth, and not something associated with the stars themselves. And this interest in refraction brings us to another of Hooke's major scientific contributions, stemming from his work on lenses, mirrors, and the nature of light.

Hooke had been fascinated by the power of lenses in extending the range of the observable world beyond the power of human eyesight ever since his time in Oxford, where he had learned astronomy from Seth Ward, the Savilian Professor. As always, Hooke wasn't just interested in using the instruments available, but in improving the technology to obtain better instruments. At that time, the telescopes used in astronomy (and elsewhere) were all refractors, with two lenses, essentially the same as the ones used by Galileo, but larger. Reflecting telescopes, in which the main light-gathering lens is replaced by a curved mirror, had been invented by the English surveyor Leonard Digges in the 1550s, and used by his son Thomas; but the knowledge had been lost, partly because they had kept it quiet (presumably for commercial reasons) and partly because the elder Digges took part in the rebellion against Queen Mary led by Sir Thomas Wyatt in 1554, and after the rebellion failed he lost his estates, and died in 1559.

The problem with refracting telescopes was that as well as gathering and focusing light, the large concave lenses produced images with coloured edges. This phenomenon, known as chromatic aberration, is now explained in terms of refraction; but in the early 1660s all that was known was that the effect could be reduced by using lenses with less curvature, which meant that they had much longer focal lengths, so that the telescopes themselves were much longer. When he was Professor of Astronomy at Gresham, Christopher Wren had a telescope 36 feet long, made for him by Richard Reeve, the best optical instrument maker in London.

An alternative way around the problem of chromatic aberration would be to develop an effective reflecting telescope, because the problem does not arise with mirrors. The Scottish mathematician

James Gregory, who had been born in 1638, came to London in 1662 with, among other things, his design for a reflecting telescope. He lacked the technical skill to make such a telescope himself, but got Reeve to make him a concave mirror of polished metal, with a focal length of 6 feet, to test in 1664. If it had worked well, this would have made a shorter, much less unwieldy telescope than Wren's 36-footer. But the tests, in which Hooke was involved, proved disappointing; the image made by the telescope lacked coloured fringes, sure enough, but it wasn't as perfectly formed as the image from a refracting telescope, because of the difficulty, at that time, of making perfectly curved mirrors. Gregory went off to Padua where he pursued his mathematical career, and Hooke stuck with refracting telescopes.

At that time, Hooke's main astronomical interest was the Moon. He studied its cratered surface using instruments like Wren's refractor, and tried to explain how the craters could have formed. In experiments in his rooms at Gresham, he found that the same kind of shallow depressions could be made by dropping round lead bullets into a sticky mixture of pipe clay and water, or by heating a pot of alabaster from beneath, so that sticky bubbles formed at the surface and burst (the same sort of thing happens with porridge). In either case, the craters looked just like miniature lunar craters. Since he could not imagine how a rain of objects could have fallen on to the Moon, Hooke plumped for the second explanation, suggesting that the Moon had (or had once had) a molten interior, and that the craters were the frozen-in remains of bursting bubbles from volcanic activity. He even suggested that the surface of the Earth, if it could be viewed from the Moon, might show similar cratering. It was only in the late twentieth century that this volcanic cratering hypothesis was proved wrong, and it was established that the Moon (and the Earth) had indeed been subjected to a rain of impacting objects long ago.

Hooke's most profound conclusion about the nature of the Moon was that its spherical shape and the arrangement of its surface features suggested that it was held together by a gravitational influence acting towards its centre, just as the Earth was. As he wrote in *Micrographia*:

I could never observe, among all the mountainous or prominent parts of the Moon . . . that any one part of it was plac'd in such a manner, that if there should be a gravitating, or attracting principle in the body of the Moon, it would make that part to fall, or be mov'd out of its visible posture.

In other words, everything on the surface of the Moon was in just the place it ought to be if it were being pulled towards the centre of the Moon.

But the Moon wasn't the only heavenly body that Hooke studied. Among his many other astronomical observations, on 9 May 1664 Hooke discovered a dark spot on the surface of Jupiter, and watched as it moved, providing the first direct evidence that Jupiter rotates on its axis.[2] To further this work, he commissioned bigger and better lenses from Reeve, while also developing his own skill as a lens grinder, inventing a new kind of grinding machine and proposing the manufacture of hollow lenses filled with a clear liquid such as alcohol or turpentine. But it was with small lenses, rather than large ones, that Hooke carried out the work which gained him widespread recognition.

Hooke used two kinds of microscope in his investigation of the world of the very small. The first consisted of nothing more than a tiny spherical lens with a huge magnifying power. These lenses were about the size of a pinhead, held in a brass mounting, and were very difficult to use because both the lens and the object being studied had to be held extremely close to the eye. But they gave the greatest magnification available at the time. The other type of microscope used a combination of lenses, rather like a small telescope, typically in a tube 6 or 7 inches long. This was much easier to use, but didn't give such great magnification. So Hooke used the compound microscope where the greatest magnification was not needed, and the single lens where attention to detail was all important. He had been interested in microscopy since his student days in Oxford, but the work which led to the publication of *Micrographia* came about because Christopher Wren was busy.

It happened like this. In 1661 Wren was asked to make a study of insects, as seen through the microscope, for the King. Although he did make several drawings, which he gave to the King, in September that year he decided that he could not spare the time to do more. So

2. In 1666, he would also be the first person to measure the rotation of Mars.

Wilkins asked Hooke to take over. The royal connection seems to have faded away as Charles became preoccupied by other things, but the upshot was that from the beginning of Hooke's work as Curator of the Royal Society he brought new drawings of what he had seen under the microscope (often subjects requested by the Fellows) along to the weekly meetings. The drawings, of course, he made himself. They were (and still are) of striking quality, showing his great skill as a draughtsman, and amazed the Fellows with their insights into an unseen world. It was the spectacular pictures of insects, in particular, that delighted and astounded readers such as Samuel Pepys when *Micrographia* was published. But with his characteristic thoroughness, when Hooke decided to make the drawings available in a book, he wanted to set them in their proper context, and to him that meant trying to explain just about everything in the known world.

The Royal Society itself was a bit ambivalent about this. *Micrographia* would only be the second book published under their imprimatur, after Evelyn's *Sylva*, and they wanted it to be a big success that would help to make the name of the Society. Hooke's drawings of insects were just the ticket; but what the Council of the Royal regarded as his more speculative hypothesizing could not, they felt, be officially endorsed by the Society. So in his dedication of the book to the Society, Hooke is careful to say that the more hypothetical parts of the book were 'not done by YOUR Directions'. Which left him free to write the book he really wanted to write, and the Royal free to bask in the reflected glory of its popular success.

In order to study anything under his microscope, Hooke had to devise ingenious arrangements of lenses and spherical containers filled with water to focus the light from the Sun on to the specimens he wanted to study. And he had to find ways to keep some of his specimens under the focus of the microscope. Studying ants proved particularly troublesome. If he killed the ants, their bodies shrivelled up, and if he stuck them down with wax or glue they twisted about so that he could not get a good view of them. The solution was to knock an ant out with brandy, 'dead drunk, so that he became moveless'.

In the book, Hooke starts with his microscopical observations of the non-living world, both natural and manmade. He shows his readers the irregular appearance of things that seem as well defined

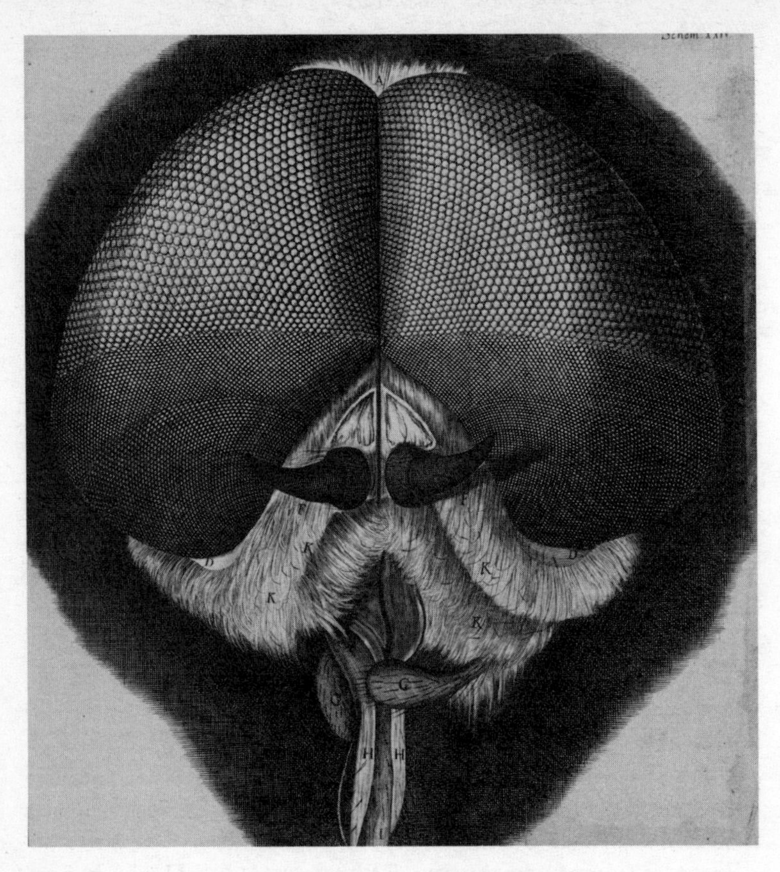

*From Robert Hooke's* Micrographia

to the unaided eye as a printed full stop, the point of a needle, and the edge of a razor. He studied crystals, and suggested that the regular external appearance of crystals must reflect a regular internal arrangement of the particles of which they were composed. He investigated the structure of cork, and found that it was arranged in tiny compartments, like a honeycomb, to which he gave the name cells – these are not what biologists now mean by cells, but they use the name in tribute to Hooke's pioneering work. He studied feathers and plants, and provided those stunning drawings of insects, including a detailed view of a fly's compound eye. Hooke explained that fossils were not stones that mimicked the appearance of living things, but were the petrified remains of creatures or plants that had indeed once been alive. Referring to the fossils now known as ammonites, he suggested that they were

> The Shells of certain Shellfishes, which, either by some Deluge, Inundation, Earthquake, or some such other means, came to be thrown to that place, and there to be filled with some kind of mud or clay, or *petrifying* water, or some other substance, which in tract of time has been settled together and hardened.

And we have already commented on the perspicacity of Hooke's astronomical speculations, also spelled out in *Micrographia*. But the one section of the book that we want to comment on in a little more detail, and which happens to be one of the longer sections, concerns Hooke's interest in light and colour. It would be this work that would draw Hooke into his famous controversy with Isaac Newton, with repercussions that affected the development of the Royal Society itself and of science in the second half of the seventeenth century.

The title of this part of the *Micrographia* is 'Of the Colours Observable in Muscovy Glass, and other Bodies'. Muscovy Glass, known today as mica, consists of many thin transparent layers of material (rather like transparent shale or slate), made of aluminium silicate. When Hooke looked at thin flakes of mica, themselves composed of even thinner layers, under the microscope, he saw a central white spot of light surrounded by a halo of coloured rings, like a circular rainbow; and he noticed how the colours moved around when the mica was squeezed. The colours were not present in the mica itself, but were

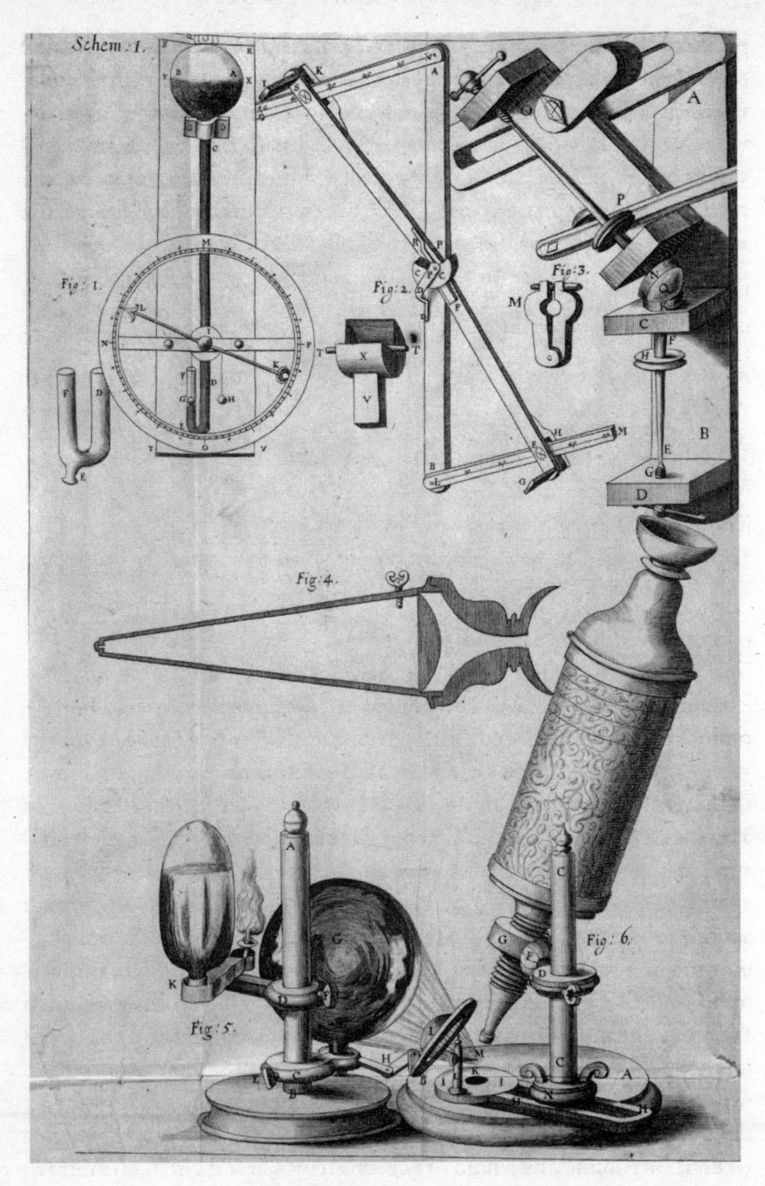


*From Robert Hooke's* Micrographia

produced where the thin pieces of transparent material were pressed together. Hooke was sufficiently intrigued to look for the same effect in other materials, and found it when two pieces of ordinary glass were pressed together, in oyster shells, and with films of sticky liquid spread over polished surfaces (the phenomenon is related to the coloured rainbow patterns seen today when a film of oil spreads across water). As Hooke put it:

> In general, wheresoever you meet with a transparent body thin enough, that is terminated by reflecting bodies of differing refractions from it, there will be a production of these pleasing and lovely colours.

And he resolved

> To examine the causes and reasons of them, and to consider, whether from these causes . . . may not be deduced the true causes of the production of all kinds of Colours.

Hooke had gone from a specific observation to the more general case, and now intended to propose a theory to account for what he had observed. He suggested that light was a form of very rapid vibration which travelled in pulses through any transparent material at an extremely high speed; but unlike many of his contemporaries (and the philosopher mathematician René Descartes, who lived from 1596 to 1650) he did not propose that the speed of light was infinite, 'for I know not one Experiment or observation that does prove it'. Hooke's ideas were always based on experiment and observation, not mere speculation.

Hooke's explanation of the production of colours started from the assumption that the natural state of undisturbed light was white, but that when it struck a refracting surface at an angle it got scrambled up in such a way as to produce colour. He thought that the front of the light pulse encountered more resistance at the surface, but that the rear of the light pulse, having had the way prepared for it by the waves in front of it, encountered less resistance. This, he suggested, made the two ends of the light pulse bend through different angles when they entered the refracting medium. On this picture, the refraction produced by the bending meant that instead of all the light waves hitting the observer's eye at right angles to the line of motion, they were twisted at various angles, and this twisting produced the impression of colour in the eye.

There's more to Hooke's discussion of colour, but it would be pointless to elaborate on it since it eventually turned out to be wrong. Strange though the idea looks to us, though, it matched the observations available to Hooke, and it was a perfectly sensible hypothesis to put forward at the time. Like all good scientific hypotheses, it set people thinking – in this case, about the nature of light. One person in particular was set thinking by this section of the *Micrographia*. Isaac Newton was, at the beginning of 1665, a 22-year-old student in Cambridge; he would receive his BA later that same year. Among his surviving papers there are seven closely written pages of notes on Hooke's book, with most emphasis on the discussion of light and colours. It would be another seven years before Newton's interest in the phenomenon described by Hooke brought him into contact with the older man; and by then Hooke's life had taken an unexpected twist.

If Hooke had been free to follow up the ideas discussed in the *Micrographia* – in particular, his ideas on gravitation and on light – the scientific revolution of the seventeenth century might have proceeded even more rapidly than it did. But within a few months of the publication of his book Hooke was affected by the first of two major events that would effectively bring an end to his great days as a scientist, though not to his work for the Royal. The first of these was the plague, which arrived in London in April 1665, and by the middle of June was causing more than a hundred deaths a week. At the end of that month, the Royal ceased its meetings and most of the Fellows left London for safer regions. The poor had no option but to stay in London, where more than a hundred thousand of them, roughly a quarter of the population, died before the Society resumed its London meetings in March 1666.

But the scientific work of the Fellows continued. Many of them moved temporarily to Oxford, where Robert Boyle was the focus of scientific society, and where they held informal meetings. Hooke stuck it out at Gresham College for a couple more weeks, partly to oversee the packing of his equipment ready for a move, and partly out of scientific curiosity about the cause and nature of the plague. His mother had died in June (not of plague), and he may have travelled to the Isle of Wight for her funeral, but by the time he left London (with deaths now running at 2,000 per week) it was too late to return to the island, which had closed its ports in an attempt to prevent the plague spreading there. Instead, Hooke joined William Petty and John Wilkins at the country estate of Lord Berkeley (one of the dilettante gentlemen Fellows) near Epsom, in Surrey.

Hooke had a string of experiments to carry out for the Royal, and was still doing work for Boyle – before leaving London, he wrote to tell Boyle of his plans, and said that if Boyle cared to send 'a catalogue' of investigations he was interested in, Hooke would 'endeavour to see them very punctually done, and to give you a faithful account of them'. When Evelyn visited the little group in Surrey in August 1665, he wrote that they were 'contriving chariots, new rigging for ships, a wheel for one to run races in, and other mechanical inventions; perhaps three such persons together were not to be found elsewhere in Europe, for parts and ingenuity'.

The 'chariots' Evelyn refers to were part of a long, time-consuming project by the Royal to improve transportation.[3] Although this was undoubtedly of practical value, and in the Baconian spirit of science improving the lot of humankind, with hindsight it is a great pity that Hooke wasn't freed to follow his deeper scientific investigations. Some of these were literally deep. Although Petty soon went to join the group in Oxford, Hooke and Wilkins found time alongside their studies of chariots to carry out a series of experiments involving lowering various objects down deep wells, measuring temperature and pressure at depth, dropping weights to test the law of acceleration, and carrying out the subterranean version of the weighing experiment, this time with one weight on a long thread down the well and the other at the surface. Hooke's interest in gravity, already stimulated by his observations of comets, seems to have grown during his time away from London. This time wasn't all spent in Surrey. In October, Hooke had to visit the Isle of Wight to sort out unresolved family business following the death of his mother, and took the opportunity

3. One was described as a saddle supported on two wheels; no details survive, but perhaps Hooke invented the bicycle.

to collect a variety of fossils and study the strata in which they were found. But we want to highlight his work on gravity, both because of its importance and because Hooke's role in this key part of the scientific revolution was largely written out of the history books.

At the second meeting of the Royal Society following the return of the Fellows after the plague, on 21 March 1666, Hooke described his experiments involving weighing objects on top of tall buildings and down deep wells, saying that he was carrying out these experiments because gravity is 'one of the most universal active principles in the world'. He was interested in finding out whether this 'attractive power', as he called it, was 'inherent in the parts of the earth', and whether 'it be magnetical, electrical, or some other nature distant from either', as well as, particularly significantly, 'to what distance the gravitating power of the earth acts'. To this end, he carried out a series of magnetic experiments, and satisfied himself that although there were similarities between the two forces, they were not the same. At the meeting on 23 May, he returned to the theme:

I have often wondered why the planets should move about the sun according to Copernicus' supposition, being not included in any solid orbs . . . nor tied to it, as their centre, by any visible strings.

He pointed out that any object under the influence of a single 'impulse' (what we would now call a force) ought to move in a straight line – this is, incidentally, close to spelling out what is now known as Newton's first law of motion – and asked where the second force required to make the planets move in curves came from. Hooke saw only two possibilities. Either the planets were moving through some fluid medium providing a drag which held them in orbit (this would only work if the fluid were denser further out from the Sun, presumably because it was colder), or the Sun must be pulling on the planets. Hooke favoured the second suggestion.

But the second cause of inflecting a direct motion into a curve may be from an attractive property of the body placed in the centre; whereby it continually endeavours to attract or draw it to itself. For if such a principle be supposed, all the phenomena of the planets seem possible to be explained by the common principle of mechanic motions; and possibly the prosecuting this speculation may give us a true hypothesis of their motion, and from some few observations, their motions may be so far brought to a certainty, that we may be able to calculate them to the greatest exactness and certainty, that can be desired . . . the phenomena of the comets as well as of the planets may [thus] be solved.

In view of the way history has treated Hooke, it's worth emphasizing that these comments were made twenty-one years before the publication of Newton's *Principia*, when Newton was still only an insignificant Fellow of Trinity College in Cambridge.

In order to show the Fellows who lacked his insight what he was talking about, Hooke hung a weight on a long wire from the ceiling of the room in which the Society met, and showed how the tendency of the weight to move in a straight line, combined with a steady inward force from the suspending wire, would make the pendulum swing around in 'orbits' that were more or less elliptical depending on the strength and direction of the original impulse. He even attached a smaller pendulum to the long one, to show how a moon could orbit round a planet while the planet was orbiting around the Sun. Nobody took much notice of these ideas at the time, probably because they did not understand what Hooke was getting at; it seems absolutely certain that Newton only became aware of the idea that planetary motions might be a product of the natural tendency of a body to move in a straight line and the attractive influence of gravity reaching out from the Sun when Hooke mentioned it to him in a letter in 1679. In Hooke's own words, written to Newton, orbital motion could be described in terms of 'A direct [straight line] motion by the tangent, and an attractive motion towards the central body'. Hooke himself (who in any case lacked the mathematical skills that Newton would bring to an understanding of planetary orbits) had failed to follow up the implications of his insight because of the dramatic change that occurred in his life as a result of the Great Fire of London, which devastated a large part of the city in September 1666.

The fire was particularly damaging because the city was largely composed of wooden buildings, crowded together in narrow streets, and there had been a long hot summer which had left them unusually dry. It raged for four days and nights, starting on 2 September 1666, and destroyed 85 per cent of the buildings within the city walls. This included St Paul's Cathedral and 84 parish churches (out of a total of 109); the Royal Exchange, the Customs House, and 44 of the halls of the livery companies; and many private houses. More than 60,000 people were not only rendered homeless, but were left with no means of earning a living, as their work places, tools, stores, and equipment were all gone.

King and Parliament, along with the City of London authorities, acted with impressive speed to resolve the crisis. With Guildhall destroyed, the Court of Aldermen which ran the City met at Gresham College, which had survived the fire, where accommodation was requisitioned for the Lord Mayor and his officers, the Sheriffs; Hooke was allowed to keep his rooms, but the Royal had to shift its meetings, initially to the lodgings of Dr Walter Pope, a stepbrother of John Wilkins, and then on a longer-term basis to Arundel House, in the Strand. The activities of the Royal Exchange also relocated to Gresham, so for the duration of the crisis Hooke was living at the heart of both the governing body and the commercial power of the City of London.

Rebuilding the city was a top priority, and also an opportunity for the architect and builders chosen for the task to make names for themselves, get rich, and do their duty as good citizens, all at the same time. Hardly surprisingly, several people produced plans for the rebuilding of the city almost as soon as the smoke from the fire had cleared; what may seem more surprising is that Hooke was among this group, although we have already seen his skill as both a draughtsman and an engineer. Thanks to his location in Gresham College he was able to present his plan (sometimes referred to as a 'model', but this means an idea, not a physical three-dimensional model) to the City authorities immediately, although he was careful to have it endorsed by the Royal. The records of their meeting on 19 September 1666 note that:

> Mr Hooke showed his model for rebuilding the city to the society, who were well pleased with it; and Sir John Laurence, late lord mayor of London, having addressed himself to the society, and expressed the present lord mayor's and aldermen's approbation of the said model, and their desire, that it might be

shewn to the King, they preferring it very much to that, which was drawn up by the surveyor of the city; the president answered, that the society would be very glad, if they or any of their members could do any service for the good of the city; and that Mr Hooke should wait upon them with his model to the King, if they thought fit to present it: which was accepted with expressions of thanks to the society.

Although details of Hooke's plan have been lost, by all accounts it was based on a grid system of streets at right angles to one another, like many modern cities of the United States. But even with the endorsement of the Royal Society and the Mayor and Aldermen of the City of London, Hooke was up against stiff opposition, including a plan offered by Christopher Wren, who, although still best known as an astronomer, had begun to make a name for himself as an architect in Oxford – he designed the Sheldonian Theatre. In the end, all of these grand plans were rejected, in favour of a straightforward rebuilding of London with essentially the same street plan as before, partly as a measure of expediency to avoid the delays that would have been caused by a grander project. On 4 October 1666, a Rebuilding Commission was set up to oversee the task, with six members, three nominated by the King and three by the City authorities. Charles appointed Christopher Wren, Hugh May, and Roger Pratt; the Aldermen appointed Peter Mills (their City Surveyor whose model they hadn't liked as much as Hooke's), Edward Jerman, and Robert Hooke, now 31 years old and in his prime.

Hooke and Wren were old friends and scientific colleagues, both Fellows of the Royal Society, and since both their grand designs for rebuilding London had been rejected they had no basis for professional jealousy in the job they faced. Hugh May was the King's Paymaster of the Works, but in 1669 Wren was appointed as Surveyor General of the works, and became May's superior, while Roger Pratt, a successful architect, soon retired to his country estate. Of the Mayor's men, Edward Jerman had considerable building experience, but died in 1668, while Peter Mills died in 1670. So Hooke (who was also sworn in as a City Surveyor in March 1667) and Wren were from the outset the dominant figures on the Commission, and within a few years were essentially running the rebuilding of London together, with little outside interference, in a partnership that lasted for three decades. It was clear from the outset which way the wind blew. Just six weeks after the fire, in a letter to Boyle in Oxford, Henry Oldenburg made no mention of the other four members of the Commission, but wrote:

> The other grand affair about the rebuilding of the Citty, is not neglected neither; Strict injunction being now issued by the Lord Mayor, in the Kings name, which done, the Survey and admeasurement of all such Foundations is to be forthwith taken in hand, and that by the care and management of Dr Wren and M. Hook: which survey is to be exactly registred; for the better stating thereafter every ones right and propriety: And then the method of building will be taken into nearer consideration, and, 'tis hoped, within a short time resolved upon.

So Robert Hooke was distracted from his scientific work, particularly in the years from 1667 to 1674, by which time most of the rebuilding work (apart from the churches) had been completed. This is not the place to go into details of his work as City Surveyor,[4] except to emphasize that in this role he was Wren's partner, not his assistant. As well as the surveying work involved at ground level, laying out the streets and buildings, deciding boundary disputes, and so on, Hooke was involved on the architectural side, designing many buildings, including the Royal College of Physicians and an unknown number of the 'Wren' churches that are still admired in the City of London today. He also designed the monument to the fire, a hollow pillar 202 feet high completed at the end of 1676. Hooke had hoped to use this tube for astronomical observations, but that proved impracticable; he did, however, carry out gravity experiments and measurements of atmospheric pressure at the top.

Hooke became a much more significant figure in London society through this work, receiving important private architectural commissions, and he also achieved the financial independence he had always craved. As a City Surveyor he initially received a salary of £100 a year, soon increased to £150. But this pales compared with his other earnings, which included a piece rate for laying out building

4. But see, for example, Lisa Jardine's biography of Christopher Wren, or *London's Leonardo*, by Jim Bennett, Michael Cooper, Michael Hunter, and Lisa Jardine.

lines for individual houses and architect's fees for his designs. Michael Cooper[5] has estimated that Hooke's work as architect and surveyor brought him a total income of about £6,000, roughly equal to the combined lifetime salaries that he received (when they were eventually paid!) from the Cutlerian and Gresham posts, and from the Royal Society. This makes a total lifetime income of some £12,000 (plus free accommodation at Gresham), and although Hooke continued to live modestly,[6] after 1667 he no longer had any need to worry about money.

While Hooke was distracted, the Royal suffered. He did not neglect his duties as Curator of Experiments completely, and still attended meetings of the Society and gave demonstrations when possible. But when he was not available, the Fellows had great difficulty finding anyone to stand in for him, and were often reduced to reading out letters they had received from other scientists. This reduced them to just the kind of gentlemen's scientific club that they had expressly intended not to be, and shows how essential Hooke was in making the Society work in the way that its founders had envisaged. We can conveniently pick up our discussion of Hooke's scientific work in 1672, when he began keeping a diary which provides great insight into his later years, city life was getting back to normal (the Royal resumed meeting at Gresham College in November the following year), and Hooke's famous dispute with Isaac Newton took place.

Taking the personal life first, it's clear from his diary and other sources (such as Aubrey and Pepys) that although he lived alone for much of his life, Hooke was a gregarious man who enjoyed the company of his friends, along with a glass of wine or brandy, or a cup or two of chocolate or coffee. Some contemporaries describe him as being irritable and difficult to get on with, and these remarks were later exaggerated by some historians to create a caricature of a grumpy old man; but such accounts do not allow for the fact that although Hooke might have had a quick temper and didn't suffer fools gladly,

5. See *London's Leonardo*.

6. At least, without ostentation. He wore plain clothes, but had them made from expensive material, and spent a relatively large amount of money on shoes. He also spent a lot on books and scientific equipment, and gave money to relatives.

such moods quickly blew over and normal friendly relations would soon be restored. It was just that people who only saw him occasionally sometimes saw the quick temper without being around long enough to see the calm after the storm.

His diary itself isn't the same kind of literary work as those of Pepys or Evelyn, but is a more telegraphic record of his day-to-day activities, like notes jotted down by a scientist while carrying out observations, ready to be written up at length later. But it is marked by complete candour, both about himself and his acquaintances. Indeed, it is so candid that in Victorian times it was regarded as unpublishable, and a full edition only appeared in print in the twentieth century. Although it is true, for example, to say that Gresham professors were required to be celibate and that Hooke lived alone for much of his life at Gresham, this hardly conveys an accurate impression of his home life. As we have mentioned, the celibacy requirement was interpreted in the strict sense that the professors were unmarried; Hooke always had a maidservant, and his diary records that he had sexual relationships with several of them. Significantly, he also seems to have stayed on good terms with them, as a friend, after they left his employment, usually to get married.

Then there was his niece, Grace, the daughter of Hooke's brother John. John had been Mayor of Newport on the Isle of Wight, and had had a successful grocery business, but by the time the diary begins he was in financial difficulties, and over the next few years Robert frequently gave him 'loans' that would never be repaid; in 1678, John Hooke committed suicide by hanging himself. At the beginning of 1672, Grace Hooke was eleven years old, and already living with Robert in London. It seems that this was a step towards an arranged marriage with Thomas Bloodworth, the son of Sir Thomas Bloodworth, a former Lord Mayor of London, but the records are unclear and there is no clue as to how or why the engagement was made. Whatever, it was broken off, and Hooke's diary for 9 July 1673 cryptically records: 'At Sir Th. Blodworths. T B gave his Discharge paper for Grace.' She remained in her uncle's household, apart from holiday visits to the Isle of Wight. Although her father was supposed to send money for her board, payments were erratic and ceased entirely in 1676. Robert Hooke paid for her education and spent far more than necessary on her clothes and presents on her birthdays. He clearly doted on her, and in June 1676, when she was just 16 and he was not quite 41, they began a sexual relationship, although at first this seems to have stopped short of full intercourse.

Probably because of his guilt about the impropriety of this relationship, Hooke sent Grace back to the Isle of Wight in 1677, but there she seems to have caught the eye of Sir Robert Holmes, the Governor of the island. According to a reconstruction of events made by Lisa Jardine, Grace had a baby by Sir Robert in 1678 (which may have been the final trigger that caused her father's suicide) before returning to London as a ruined woman. If so, Hooke took her back into his household not merely for selfish reasons but out of compassion, making her his official housekeeper (and unofficial mistress) and giving her the security that she could not otherwise have expected, now that she had no prospect of a good marriage.

Although Grace caused Hooke mental agonies from time to time by her understandable interest in young men her own age, she stayed with him long after she could have left to make her own way in the world, and the relationship does not seem to have been entirely one-sided, although Hooke clearly felt more deeply for her than she did for him. When she died in 1687, at the age of only 27, he was devastated, and never really recovered. According to Richard Waller, who edited Hooke's *Posthumous Works*, he was 'observ'd from that time to grow less active, more Melancholly and Cynical'.

Hooke himself was 52 in 1687, and even while Grace was alive he had had an alarming (to modern eyes) habit of dosing himself with quack medicines and patent remedies, for a series of real or imagined illnesses, usually to do with his digestion. The very first entry in his diary reads, 'Drank iron and mercury', and its entries record doses of such divers substances as (in alphabetical order), Aldersgate cordial, aloes, Dulwich water, Epsom water, flowers of sulphur, hagiox, laudanum, lignum vitae, North Hall water, resin of jalop, rhubarb, sal ammoniac, senna, wormwood, and metals in any drinkable compound forms. The references to self-medication in the diary are so common, indeed, that an entry like that for 3 August 1673, which simply reads, 'At Garways in the evening. Took noe physik', is the startling exception to the rule. You don't need to know what all these substances were to realize that Hooke was poisoning himself, and it is hard from this distance in time to know to what extent problems such as 'that viscous Slime that hath soe tormented me in my stomack and gutts' were caused by the illness or the 'cure'. All this got worse after Grace died.

Scientists seldom do great original work after the age of 40 in any case, but the problems caused by ill health and the potions he took would not have helped; and as we have seen, what might have been some of Hooke's most productive scientific years, bringing to fruition his work on optics and gravity, were lost to science when he became City Surveyor after the Great Fire. It was just at the time the city itself began to get back to normal, and just before the Royal returned to Gresham College, that Hooke, in his 37th year, first encountered his Nemesis. At the end of 1671 Isaac Barrow, who was an eminent Cambridge mathematician, as well as having been Hooke's predecessor as Gresham Professor of Geometry and an early Fellow of the Royal, drew the attention of the Society to the work of his protégé Isaac Newton, by now the Lucasian Professor of Mathematics in Cambridge. Barrow showed the Society a reflecting telescope designed and built by Newton, which was a different design from the Gregorian telescope and seemed to get around many of the problems with earlier instruments.[7] The telescope was given to a committee consisting of Brouncker, Hooke, Moray, Neile, and Wren to be tested, and they were sufficiently impressed for the Society to accept the instrument as a gift on 11 January 1672, with Newton being elected as a Fellow at the same meeting.

Hooke wrote to Newton about his own work with reflectors, and started work on a new instrument of his own, more than a little irritated to see Newton receiving accolades for something which he might have done himself, if he had not been so busy for the past ten years. But the irritation grew to anger when Newton, having been asked by the Royal what else he had been working on, sent the Society a letter setting out his experiments and ideas on light and colour, which was read to the Society on 8 February. The problem was

7. In fact, Newton's telescope soon proved as impractical at the time as Gregory's, as the mirror was difficult to shape and easy to tarnish.

that although Newton's work clearly jumped off from Hooke's work described in *Micrographia*, the young upstart (as Hooke saw him) failed to give due credit to the older, established scientist. Hooke was always particularly touchy about getting the credit he deserved, largely because he was aware of his position as a paid servant of the Society, and felt (rightly) that some of the Fellows did not regard him as a proper gentleman, and didn't fully appreciate his contributions. His relationship with Oldenburg was particularly troubled, because of Oldenburg's habit of passing on sensitive news about discoveries and inventions to third parties without permission, and sometimes without credit; since Wren, who scarcely ever had a bad word to say about anyone, also referred to Oldenburg's 'disingenuity and Breach of Trust',[8] this was clearly not just paranoia on Hooke's part.

Newton's letter (we would now call it a scientific paper) was copied and given to three Fellows to assess – Robert Boyle, Seth Ward, and Hooke. The first two seem to have made no written report on the paper, although presumably they commented on it verbally, but Hooke offered a detailed critique. Newton's key suggestion was that white light is a mixture of all the colours of the rainbow, which are separated by being bent by different amounts as they pass through a prism; he also suggested that light was made up of a stream of tiny particles, rather than travelling as a wave. In his critique, Hooke recapitulated his own theory of light and colours, and made the entirely reasonable point that although Newton's model might provide *an* explanation for the phenomena he observed, it was not the *only* possible explanation. But, as was his wont, he got rather carried away with his criticism:

> I do not therefore see any absolute necessity to believe his theory demonstrated, since I can assure Mr Newton, I cannot only salve all the Phenomena of Light and colours by the Hypothesis I have formerly printed . . . but by two or three other . . . Nor would I be understood to have said all this against his theory as it is an hypothesis, for I . . . esteem it very subtill and ingenious . . . but I cannot think it to be the only hypothesis; not so certain as mathematicall Demonstrations.

8. Quoted by 'Espinasse.

In other words, Hooke saw Newton's idea as an acceptable working hypothesis, which might be tested by further experiments, but not as a proven theory. History shows that this approach was correct, and that he and Newton were each half right – light does travel as a wave, as Hooke suggested, but white light is a mixture of colours, as Newton suggested.

Hooke presented his critique to the meeting of the Royal on 15 February 1672, and Oldenburg sent a copy to Newton; but the Society decided to publish Newton's paper in the *Philosophical Transactions* without Hooke's comments. You might think that Newton would have been pleased; but he was a very strange individual (as we shall see in the next chapter) who was neurotically sensitive to criticism of any kind. He didn't immediately take offence at Hooke's comments, though, and was only stirred into action by Oldenburg, who mischievously tried to use Newton to make trouble for Hooke. At Oldenburg's urging, in June Newton sent a strongly worded response to Hooke's critique, taking Hooke to task for allegedly failing to understand Newton's arguments, and telling Hooke that 'it is not for one man to prescribe Rules to ye studies of another, especially not without understanding the grounds on wch he proceeds'. The existence of several drafts of the letter show how Newton became increasingly angry as he worked at it, starting out reasonably calmly but ending up with a version which turned into a personal attack on Hooke. And Hooke had to sit in the audience at the Royal while Oldenburg read out the juiciest bits of the letter at the 12 June meeting. But it's worth pointing out that Newton wasn't entirely hostile at this time; although he didn't agree with Hooke's wave model of light, he did suggest that it might be adapted to explain his own observations of refraction if different colours corresponded to waves with different 'depths or bignesses', which were refracted by different amounts. But in the face of criticism from Hooke and others, after this contribution Newton decided that he had no time for any more such arguments, and that to avoid criticism he would keep his ideas to himself in future.

This was the origin of the acrimony between Hooke and Newton (always more intense on Newton's side) which simmered over the next few years, and eventually burst into flame. In the meantime, though, among his many activities as Curator of Experiments,

Cutlerian Lecturer, Gresham Professor, and City Surveyor, Hooke was involved in the only other major scientific dispute of his life, and once again Oldenburg was the catalyst. In 1675, Christiaan Huygens sent news to the Royal Society of his invention of a spring-driven watch. The design seemed strikingly similar to the one Hooke had suggested to the Royal in the early 1660s. It turned out that Huygens, as a foreign national, had offered Oldenburg the English patent rights on the watch (not realizing that Oldenburg, also a foreign national, was not allowed to hold an English patent), so Hooke immediately suspected collusion between them to steal Hooke's idea. In fact, it turned out that Huygens had learned about Hooke's design in a letter from Robert Moray, in 1665. But Huygens now had a working watch, and Hooke did not. He enlisted the help of the watch maker Thomas Tompion to put his idea into practical form, producing a watch that he presented to Charles II on 17 May, bearing the inscription 'R. HOOK invenit 1658. T. TOMPION fecit 1675'. This led to a kind of arms race between Hooke and Huygens to produce better watches, and also (whatever the rights and wrongs of the affair) to a deepening of the rift between Hooke and Oldenburg. Fate soon gave Oldenburg his chance for a renewed attack on Hooke by proxy.

In February and March, Isaac Newton had been in London on other business, and took the opportunity to attend three meetings of the Royal Society. He must have been surprised and pleased to hear Hooke speaking at the third of those meetings in support of Newton's experiments with prisms, in response to a letter from a scientist of the old school, Francis Linus, who had claimed that the separation of colours by refraction did not happen. As a good scientist, Hooke was bound to accept the experimental evidence for the separation of colours, even though he did not agree with Newton's interpretation of *why* the separation happened.

A week later, after Newton had returned to Cambridge, Hooke described a modified version of his own ideas about colour to the Society, taking on board Newton's suggestion from the summer of 1672, and a week after that he described the discovery of a new phenomenon that he called 'inflexion'. This is the process we now call diffraction, which happens when waves bend round the corner of an obstacle – you can see the effect at work with ripples on a pond, or waves in the sea passing the end of a jetty. By placing the edge of a razor across a beam of light, Hooke discovered that the shadow behind the razor did not have a sharp edge between black and white, but that a little light leaked into the edge of the shadow. Once this process was properly understood it would be taken as definitive proof that light travels as a wave; but that did not happen in either Hooke's or Newton's lifetime. In fact, this discovery also dated back to 1672, when Hooke had mentioned it in a letter to Brouncker, but this was its first official, public airing. Newton commented that the effect was simply a new kind of refraction; he was wrong, but in any case, as Hooke replied, 'though it should be but a new kind of refraction, yet it was a *new one*'.

Newton had been so well received at the Royal that he decided to come out of his shell and communicate with them once again; the immediate upshot was a very long paper on light that Oldenburg read to the Society in two parts, on 9 and 16 December 1675. The second part of the paper included an attack on Hooke's explanation of diffraction, and pointed out that the effect had already been discovered by the Italian Francesco Grimaldi, who actually gave it the name diffraction. The discovery was reported in a book called *Physico-Mathesis de Lumine, Coloribus, et Iride*, published in 1665, two years after Grimaldi had died. In his paper, Newton wrote, 'I make no question but Mr Hook was the Author too,' meaning that he was sure that Hooke had made the discovery independently. But Oldenburg struck this line out, and did not read it at the meeting, giving the impression that Newton thought that Hooke had simply lifted the idea from Grimaldi. Hooke's quick temper boiled over, and he responded to the effect that Newton was in no position to accuse Hooke of plagiarism, since almost everything in Newton's paper 'was contained in [Hooke's] *Micrographia*, which Mr Newton had only carried farther in some particulars'. By the time Hooke had (equally characteristically) calmed down, Oldenburg had written to Newton with an exaggerated version of Hooke's comments.

This led to a furious response from Newton, as Oldenburg had no doubt intended, saying that Hooke had done nothing original and that his ideas about colour were largely taken from Descartes. Oldenburg read Newton's letter to the Society on 20 January 1676. Hooke, who had not been forewarned, was stunned, but realized what was going on. In his diary that evening he wrote of the meeting:

Mr Auberys papers Read and some of Mr Newtons. A letter also of Mr Newtons seeming to quarrel from Oldenburg fals suggestions . . . Wrot letter to Mr Newton about Oldenburg kindle Cole.

It was this letter, an attempt by Hooke to put out the flames fed by Oldenburg's kindling of the coal, that produced one of the most famous, and misunderstood, phrases in science. Hooke wrote in measured, conciliatory tones to Newton:

I judge you have gone farther in that affair [the study of light] much than I did, and . . . as I judge you cannot meet with any subject more worthy your contemplation, so I believe the subject cannot meet with a fitter and more able person to enquire into it than yourself, who are every way fitted to compleat, rectify and reform what were the sentiments of my younger studies, which I designed to have done somewhat at myself, if my other more troublesome employments would have permitted, though I am sufficiently sensible it would have been with abilities much inferior to yours. Your Designes and myne I suppose aim both at the same thing wch is the Discovery of truth, and I suppose we can both endure to hear objections, so as they come not in a manner of open hostility, and have minds equally inclined to yield to the plainest deductions of reason from experiment. If, therefore, you will please to correspond about such matters by private letter I shall very gladly embrace it; and when I shall have the happiness to peruse your excellent discourse (which I can as yet understand nothing more of by hearing it cursorily read) I shall if it be not ungrateful to send you freely my objections, if I have any, or my concurrences, if I am convinced, which is the more likely. This way of contending, I believe to be the more philosophicall of the two, for though I confess the collision of two hard-to-yield contenders may produce light yet if they be put together by the ears by other's hands and incentives,[9] it will produce rather ill concomitant heat, which served for no other use but . . . kindle-coal.

This is not only reasonable, but sums up the essence of the scientific approach which the Royal Society was all about – the discovery of

9. He is referring to Oldenburg's mischief-making.

truth, and a willingness to listen to different ideas, which could be tested by experiment. Unfortunately, Newton could never 'endure to hear objections', no matter how well founded, and although his reply to Hooke reads in the same spirit of conciliation, the seeds of further disputes with Hooke had already been sown. Even in this letter, although it starts pleasantly enough, there is a sting in the tail which has been interpreted as an intended snub to Hooke:

At ye reading of your letter I was exceedingly well pleased & satisfied wth our generous freedom, & think you have done what becomes a true Philosophical spirit. There is nothing wch I desire to avoyde in matters of Philosophy more than contention, nor any kind of contention more than one in print: & therefore I gladly embrace your proposal of a private correspondence. What's done before many witnesses is seldome wthout some further concern than that for truth: but what passes between friends in private usually deserves ye name of consultation rather than contest, & so I hope it will prove between you & me. Your animadversions will be therefore very welcome to me: for though I was formerly tired with this subject, & have not yet nor I beleive ever shall recover so much love for it as to delight in spending time about it; yet to have at once in short ye strongest or most pertinent Objections that may be made, I could really desire, & know no man better able to furnish me with them than your self. In this you will oblige me. And if there be any thing els in my papers in wch you apprehend I have assumed too much, or not done you right, if you please to reserve your sentiments of it for a private letter, I hope you will find also that I am not so much in love with philosophical productions but yt I can make them yeild to equity & friendship. But, in ye meane time you defer too much to my ability for searching into this subject. What Des-Cartes did was a good step. You have added much in several ways, & especially in taking ye colours of thin plates into philosophical consideration. If I have seen farther, it is by standing on the shoulders of Giants.

Part of the subtext of this letter, courteous though it is, is that Newton (like all of us!) would prefer Hooke to keep any criticisms private, and not to tell the world about his mistakes. The reference to Descartes can be read as a gratuitous reminder of Newton's earlier claim that Hooke hadn't really done anything original, and the phrase about standing on the shoulders of Giants is also more a dig at Hooke than modesty by Newton, and in spite of the popular myth has nothing to do with Newton's work on gravity, and does not refer back to Galileo, Kepler, and Copernicus. It actually refers back to Roman times, and Newton seems to be implying that he may have built on the work of great men of the past but that (by implication) Hooke is not one of the greats. It has even been suggested that this was a deliberate reference to Hooke's small stature, although the case can never be proven. What is beyond doubt, however, is that after this exchange, even though the two men continued to correspond on various topics, Newton (as we shall see) waited until Hooke was dead before publishing anything else on light and colour, and that by so doing he managed to ensure that the phenomenon of coloured patterns associated with thin plates, discovered by Hooke, is known to this day as 'Newton's rings'.

Hooke's scientific activities continued throughout the 1670s and beyond, alongside his many other activities, and his ideas about earthquakes and the age of the Earth, presented in the published version of his Cutlerian lectures, were particularly striking and ahead of their time. But just because they were ahead of their time, they had little influence on the development of science in the seventeenth century. As far as that story is concerned, it is natural now to shift the spotlight to Isaac Newton, who, for all his faults, essentially completed the task of spelling out what science is all about that had been begun by William Gilbert and Francis Bacon, and in so doing raised the status of the Royal Society to unprecedented heights. Hooke still comes into that part of the story, as a catalyst for Newton's greatest work; he was only 41 in 1676, and still had more than a quarter of a century of his full and satisfying life ahead of him. This was the year that his sexual relationship with Grace began, and in his public life he continued to work as a surveyor for the City and privately as an architect, lectured and published on various aspects of science, and remained Curator of Experiments at the Royal (after 1683, with two other curators to ease the burden on him). After Oldenburg died in 1677, Hooke was elected as one of the Secretaries and to the Council of the Royal Society, which must have been particularly pleasing. At the same time, the ageing Brouncker was replaced as President by Sir Joseph Williamson (the then Secretary of State), and Christopher

Wren became Vice-President. This group of close friends was at the heart of a concerted attempt to revive the early enthusiasm of the Royal, and to return the Society to its roots in experimental science over the next few years. Hooke lost both the Secretaryship and his place on Council in 1682, but was re-elected to the Council in 1684, and sat on it in all for a further thirteen years.[10] He made his last recorded contribution to a Royal Society meeting on 19 June 1702.

In his last years Hooke lived alone except for a maidservant, and although he had become rich from his surveying work, he was so afraid that his money would not see him out that he became miserly and resented spending money even on decent food. He died, after a long illness, on 3 March 1703, a few months short of his 68th birthday. Hooke had always said that he intended to leave his money to the Royal Society to build a library and endow a lectureship, but he never got round to making a will (at least, none was ever found), and the £8,000 fortune left in his rooms in a great iron chest went to distant relatives on his mother's side of the family. But his funeral was held in some style, and attended by every available Fellow of the Royal Society.

Robert Hooke was the best experimental scientist of his time, as witnessed by the fact that both Robert Boyle and the Royal Society picked him out for the job; he was the leading microscopist of the seventeenth century, and an astronomer second to none; his understanding of earthquakes, fossils, and the history of the Earth would not be surpassed for more than a hundred years. As an architect, he wasn't quite in the same league as Christopher Wren, it is true; and as a mathematician, he was certainly not in the same league as Isaac Newton (but then, who was?).

It was this last failing (if that is not too strong a word), coupled with Newton's personal animosity towards Hooke, that allowed Hooke's perceived status to decline in the century that followed his death, as science became mathematical. Indeed it is only in the past few years, three centuries after his death, that Hooke's enormous contribution to the revolution in science which happened in England in the seven-

10. During this time, incidentally, Hooke engaged in correspondence with Thomas Newcomen about his atmospheric engine, the forerunner of the steam engine.

teenth century has been fully appreciated. Nevertheless, the final step in establishing the basis of science as we know it was the one taken by Newton, who made it mathematical. It's there in the very title of his epic book, which translates as *The Mathematical Principles of Natural Philosophy*; the emphasis is on the mathematical principles, and that is what makes modern science so powerful. But to see just how a farmer's son from Lincolnshire completed the revolution in science, and what he did next, we will have to go back to his birth on Christmas Day 1642.

# BOOK THREE

# Coming of Age

*Sir Isaac Newton*

# 7

# Tribulation and Triumph

Isaac Newton played two key roles in establishing the position of the Royal Society at the heart of the scientific revolution. His great book the *Principia* was the most important single publication in the history of science, and its association with the Royal helped to establish the Society as *the* place where great issues in science were debated and published. And in later life, in a surprising turnaround from his early shyness and reticence at getting involved in public debate, Newton became the President of the Royal Society, ruling it with a firm hand and completing the process whereby a gentlemen's club became a pillar of the establishment. This might never have happened, for, after the initial burst of enthusiasm which saw the Society established, it went through many difficulties in the last decades of the seventeenth century, quite apart from the problems caused by plague and fire, before finally achieving some sort of financial security and a core Fellowship of real scientists.

We don't wish to go into any great detail about these tribulations, but a little background may help to put Newton's contribution into perspective. The problems stemmed from the fact that in spite of the Royal Charter and the King's initial enthusiasm for the project, the Royal never did receive any kind of official funding, or even a building to call its own. Although it had never been the intention of the founders, in the long run this may have been a good thing, since it meant that the Society was always independent, and seen to be independent. The French Académie des Sciences, for example, was always a government institution, which meant that it had funds, a building, and could pay salaries for scientists; but the quid pro quo was that the Académie, and those scientists, had obligations to the government, and had to carry out work for the government, rather than being free spirits. There are good and bad points about both systems, but at least the survival of the Royal meant that there were two systems in operation side by side.[1]

One way of getting an idea of the ups and downs of the Royal Society from its foundation in the pre-charter meeting of 28 November 1660 to the end of the seventeenth century is to look at the changing nature of its membership. There were only 551 Fellows elected during that period, and Michael Hunter has looked at their backgrounds, and what these reveal about the fortunes of the Society itself. The most notable point is that the earliest membership of the Society did not include every competent scientist (not even all the scientists resident in London), while many of the early Fellows were by no stretch of the imagination competent scientists. In order to get started the Society needed money (from subscriptions and gifts) and prestige, and it got both by encouraging members of the aristocracy to join – indeed, Privy Councillors and aristocrats above the rank of baron were admitted without any investigation of their scientific worth. Government ministers such as the Lord Chancellor and the Lord Privy Seal followed the example of the King, his brother the Duke of York, and Prince Rupert in becoming Fellows, and there was a fashion for science in the 1660s. Hunter quotes a letter describing how while the King and his party were based at Salisbury during the plague year of 1665, they would amuse themselves in the evening by attending lectures given by Sir William Petty and other scientists 'upon something that nobody understands but [the lecturer]', and a slightly later contemporary criticizes the Royal for 'so readily admitting all persons into their Society, who will pay the Duties of the house, though they know not the terms of Philosophy'. But there lay the rub – the Society *had* to accept those who would pay the 'duties of the house', simply in order to survive.

The Society was actually run by a small inner core of Fellows, no more than twenty at any one time, although their membership changed as time passed. These were aided (not just financially) by a larger group of Fellows who took some part in the activities of the Society

1. The Académie was founded four years after the Royal Society received its Charter.

and at least paid their subscriptions promptly. As well as the subscription of a shilling a week, Fellows were asked from time to time to make contributions towards specific costs, and even Fellows on the periphery of the scientific activities of the Society helped in this way to keep it going. Judging from Hooke's diary, in the first decades of its existence the actual numbers attending the weekly meetings were usually about twenty, but often as low as a dozen or so; but this wasn't made up of the same faces every week, the hard core of the Fellowship, but rather drew from a larger group of Fellows who each attended some meetings although not every one. Hunter's analysis shows that the active interest of members in the Society reached a peak in 1665, shortly before the outbreak of the plague, when 104 Fellows were actively involved in one way or another. By 1672, the active membership was down to 80, and a slow decline took the figure down to 75 in 1685 and a low of 55 in 1692, followed by a recovery to 76 in 1700.

During the period up to the end of the seventeenth century, new Fellows (as might be expected) tended to be active for a couple of years and then lose interest in the Society; but against this background (again, hardly surprisingly) the Fellows elected in the first two years of the Society's existence (before it became Royal) were more active and maintained their interest for longer. After all, they were the people who had brought the Society into being. As late as 1672, just over half of the active core of the Society was still made up of Fellows elected within a year of the 28 November 1660 meeting. In spite of the deaths of John Wilkins in 1672 and Robert Moray in 1673, they still made up nearly 30 per cent of the core membership in 1685, although old age and death then took its toll. It is no coincidence that the low point in the Society's activity occurred at the end of that decade and the beginning of the 1690s.

On the financial side, there were difficulties from the outset. Even before the outbreak of the plague, the arrears of subscriptions amounted to more than £300, while by the end of the 1660s the figure had grown above £1,000. A few of the worst offenders were expelled from the Society, but most were simply urged to pay up – after all, if they were expelled there would be no hope of getting any money from them. Strenuous efforts were made to restore the Society's fortunes in the mid-1670s, both in terms of its scientific activity (which had suffered ever since Hooke got distracted by architecture) and in terms of its finances, with legal action taken to recover some of the debts and members asked to sign a new, legally binding promise to pay subscriptions promptly in future. But by the 1680s there were real fears that the Society might fail. It was saved by an influx of new blood that included John Flamsteed (who would become the first Astronomer Royal) and Edmond Halley (of whom more in the next chapter). It may also have helped that Oldenburg, not always the easiest of men to work with, died in 1677. When Christopher Wren became President of the Royal in 1680, he took a firm grip on its affairs, which resulted in the extraction of a great deal of the arrears of subscriptions from some Fellows, and the eventual expulsion of more than sixty Fellows, a purge which dramatically altered the character of the Society. Wren also changed the policy of the Society towards new Fellows, ensuring that successful candidates for membership had to be capable of making a genuine contribution to its scientific affairs.

The Society also had something of a windfall in 1682. Soon after granting the Society its Royal Charter, Charles II had given it Chelsea College as a site on which to build its headquarters. Legal wrangling about the terms of this arrangement, and the lack of funds to build anything, had made this gift useless (actually worse than useless, since the legal wrangling cost money), but in 1682 the King bought the land back, for the sum of £1,300, to be the home for a royal hospital (now the home of the famous Chelsea Pensioners). It was this background that enabled the Society, at last, to employ other curators to supplement the work of Robert Hooke.

Unfortunately, the revival in the Society's fortunes seems to have induced some complacency. Now that it had funds, it invested them unwisely, while some Fellows (perhaps lulled into thinking that the Society was no longer so dependent upon them) again became unreliable at paying their subscriptions. As we shall see, Edmond Halley, by then employed by the Society as a Clerk, was one of the people to suffer financially as a consequence, and in 1692 the total expenditure of the Society was a mere £37 15s 10d. We can be sure that they would have spent more if they had had more to spend – a decade earlier, the equivalent figure had been £198 14s 8d. These difficulties were followed in turn by another bout of reform in the late 1690s, with an amnesty for outstanding arrears being offered in exchange (as usual) for a promise to pay promptly in future, and the introduction of an improved accounting system. Once again, Fellows were urged to provide experiments for meetings, and by 1700 there were 41 core members of the Society. It is against this background of a society struggling to survive, with modest highs punctuated by deep lows, that the contribution of Isaac Newton to the Royal, first as a scientist and then as its President, can be understood.

Although Newton made an immense contribution to the revolution in science, and was undoubtedly a great scientist, in popular mythology this has led to an exaggerated account of his importance. The myth goes something like this. Isaac Newton was the greatest scientist of all time. Although the first steps had been taken by Galileo Galilei (William Gilbert gets ignored in the popular myth), it was Newton who perfected the modern technique of scientific investigation, in which ideas are tested and refined by comparison with experiments, instead of being plucked out of the air as more or less inspired speculations. Although Newton's direct discoveries and inventions in science were outstanding – he discovered the law of gravity and the laws of motion, made important advances in the understanding of light, designed and built with his own hands a new kind of telescope, *and* invented the mathematical technique of calculus – it was his way of doing science that both made all this possible and transformed the investigation of the natural world.

It's all true, up to a point, and it is amazing that it was all done by one man. But Newton's real importance derives from the way in which he pulled together all the various bits and pieces of the new philosophy into a coherent whole. If he had not existed, however, this would have happened anyway, certainly within a generation, and the contribution of others (notably Hooke) might have been better remembered. Alexander Pope's famous couplet

> Nature and Nature's laws lay hid in night:
> God said 'Let Newton be' and all was light

may be clever, but it is far from accurate.

Newton himself was quite clear about the importance of the new approach to science, and of the way it differed from the approach of many of his contemporaries and immediate predecessors. He once wrote – in a letter to the French Jesuit Gaston Pardies, in connection with the public attention stirred by Newton's theory of light and colours – that

> The best and safest way of philosophizing seems to be, first to enquire diligently into the properties of things, and to establish those properties by experiences [experiments] and then to proceed slowly to hypotheses for the explanation of them. For hypotheses should be employed only in explaining the properties of things, but not assumed in determining them; unless so far as they may furnish experiments.

This is at the heart of modern science. If your favourite idea about the nature of the world does not match up with the results of experiment, then it is wrong. But it is also at the heart of what William Gilbert did, and wrote, more than half a century before Newton.

The 'Newtonian' approach to science is now so deeply ingrained in anyone who has studied science at all that it is hard to realize just what a big step forward it was in the seventeenth century. But even in Newton's day, in spite of the pioneering efforts of Gilbert, Harvey, and a few others, there were still people who would argue, when he put forward his ideas about light, that these ideas were so obviously wrong that there was no need to carry out experiments to test them. It wasn't Newton alone who carried the day, but it is essentially true to say that it was after Newton that the new philosophy became established as *the* way to do science.

Largely thanks to this scientific approach to problems, and to his towering intellect, Newton finally removed the need to invoke magic or the supernatural to explain the workings of the Universe. Although Nicolaus Copernicus had suggested that the Earth moved around the Sun, and Galileo had gathered a wealth of evidence that it did so, before Newton nobody knew what held the planets in their orbits, or what kept the stars in their places in the sky. It was Newton who showed that the Universe works in accordance with precise rules, or laws, and that the motion of the planets and comets (and, by implication, even the stars) could be explained by the same laws that applied to the fall of an apple or the flight of a cannonball here on Earth. His law of gravitation and laws of motion are *universal* laws that apply everywhere, and at all times.

The lives of Galileo and Newton together span almost exactly the period of the scientific revolution. Galileo was born in 1564 and died in 1642. Newton was born within twelve months of the death of Galileo, and died in 1727. In 1564, the world was a mysterious place, what passed for science was the ancient lore of Aristotle (purely 'philosophical' ideas unsullied by any contact with something as basic as an experiment, but enshrined by centuries of tradition), and ideas about the Universe at large were mainly superstitious nonsense. In 1727, the modern image of the Universe as a great machine, ticking away as steadily and as predictably as clockwork, was firmly established (at least for the educated and literate members of society), and in a sense the Universe had been tamed.

But the image you should carry away is not one of a modern wristwatch, unobtrusively marking the passage of time, but rather of a great cathedral clock of the early eighteenth century, driven by the swing of a huge pendulum (itself following laws first worked out by Galileo), with many interconnecting cogs and gearwheels, working together not only to mark the passage of time but to drive a complicated mechanism controlling the motion of elaborate tableaux including moving figures of the saints, and striking a variety of bells at the appointed hours and quarters.

Newton's work made it clear that, like the complexity of those clocks, driven by a simple swinging pendulum, the Universe obeys simple laws which are intelligible to the human mind, and that it is these simple laws that interact to produce the complexity of the world we see around us. And that realization, plus the experimental method, is the underpinning of all modern science.

Isaac Newton came from a line of successful, upwardly mobile farmers, who had no tradition of learning whatsoever – neither his father nor his paternal grandfather could read or even write their own name. The grandfather, Robert Newton, was born around 1570, and as well as inheriting land at Woolsthorpe, in Lincolnshire, he prospered sufficiently to add to his property by purchasing the manor of Woolsthorpe in 1623, which carried with it the improved social status of Lord of the Manor of Woolsthorpe. It is quite likely that this social elevation was an important factor in enabling Robert's oldest surviving son, Isaac, to become betrothed to Hannah Ayscough in 1639.

The Ayscoughs were the social superiors of the Newtons (Hannah's father, James, was described as a 'gentleman' in official documents), and brought to the prospective marriage property worth £50 a year, while Robert Newton made Isaac the heir to his entire Woolsthorpe property, including the Lordship of the Manor.

Far from being illiterate, the Ayscoughs were well educated, and Hannah's brother William was a Cambridge graduate and clergyman who took up a good living at a nearby village, Burton Coggles, in 1642. That was the year in which Hannah and Isaac Newton senior married, in April, some six months after the death of Robert Newton. In another six months, Isaac himself also died, leaving his widow comfortably provided for, but pregnant with a baby that was born, prematurely, on Christmas Day, and christened Isaac in memory of his father.

At least, it was Christmas Day in England. The Gregorian calendar had been introduced in Catholic countries in 1582, but in Britain (suspicious of popish trickery) the reform was not applied until 1752. At the time 'our' Isaac Newton was born, the continental calendar was ten days ahead of the Julian calendar still in use in England, so it was already 4 January. That is why it is really cheating to say that Newton was born in the same year that Galileo died (on 8 January 1642 according to the Gregorian calendar); the 'coincidence' depends on using dates from two different calendar systems!

Another calendrical complication arises, as we have mentioned, from the fact that in England at that time the legal date of New Year was 25 March. This is still enshrined in the tax year, which now begins in England early in April (thanks to the Gregorian calendar reform). As ever, we shall ignore this complication and give dates as if the year began on 1 January, in line with modern practice.

Whatever the date, when Isaac was born he was so small that, as he used to recount gleefully in later life, he could have been fitted into a quart pot (it is not clear if he ever *was* treated to such an indignity), and he was not expected to live. But he survived and flourished until he was three years old, when his life was turned upside down.

Many other lives were being turned upside down around that time, by the English Civil War. Although the repercussions of these upheavals would affect Newton in later life, the turmoil of the Civil War itself seems to have largely passed Woolsthorpe by. Newton's upheaval in 1645 was much more personal – his mother remarried, and he was sent to live with his maternal grandparents.

Hannah's second marriage was a businesslike affair (as, indeed, her first had been). Barnabas Smith, her new husband, was 63 years old, wealthy, and had buried his first wife in June 1644. He was the Rector of North Witham, less than two miles from Woolsthorpe, and had looked around for a suitable woman to replace his loss, settling on Mrs Hannah Newton and approaching her through an intermediary. On the advice of her brother, Hannah accepted, on condition that her new husband immediately settled a piece of land on young Isaac,[2] and she went off to live with Barnabas Smith in the rectory, where she bore him two daughters and a son before he died at the age of 71 in 1653.

Hannah, her wealth considerably enhanced by the inheritance from her second husband, then moved back to Woolsthorpe, and reclaimed Isaac for her own household; but for eight crucial years he had been living as a solitary child with his elderly grandparents. Since they had married in 1609, by 1645 they must have been about the same age as Hannah's new husband.

This upheaval is usually presented as an entirely bad experience for Isaac, and it is hardly probable that he enjoyed those years. The likely psychological consequences of having been taken away from his mother after three years as the only child of a doting single parent are obvious enough, and his relationship with his grandfather (the 'gentleman') was such that he never once mentioned him in his writings, while Isaac himself was not mentioned in the grandfather's will. But although he was lonely and probably unloved (a list of 'sins' that he wrote down a few years later includes 'threatening my father and mother Smith to burn them and the house over them'), the Ayscough experience provided him with one thing that he might not otherwise have had – an education.

2. Newton came into this inheritance when he was 21; together with his later inheritance from his mother, it meant that he never had any financial worries as an adult.

There is no reason to think that Isaac Newton senior, had he lived, would have thought it worth having his son, destined to be a farmer, educated at all. It might have occurred to Hannah, had she stayed a lonely widow – but, equally likely, she might have wanted the boy to learn about the farm as soon as possible. To the elder Ayscoughs, though, it was automatic to send the boy to day school and ensure that he was set on the road to a better education than a farmer's son might expect (besides, school kept him out of the house).

Isaac returned to his mother's home in 1653, but in less than two years, at the age of 12, he was sent away to study at the Grammar School in Grantham, some 5 miles (8 kilometres) away, where he lodged with the family of an apothecary, Mr Clark. The school was a good one which provided a grounding in the Classics – mainly Latin, some Greek, and a little mathematics. It taught nothing else, of course; but the Latin would be invaluable to Newton in his scientific career, because it was then still the universal language used for communication by academics across Europe, and in which important books were written. The school itself had a notable old boy, the poet and philosopher Henry More (1614–87), who by 1654 was a Senior Fellow of Christ's College, in Cambridge; and the Clarks were no ordinary apothecary's family, for Mrs Clark had a brother, Humphrey Babington, who was a Fellow of Trinity College, but spent most of his time at Boothby Pagnell, near Grantham, where he was rector. There were people in young Isaac's circle of acquaintances in Grantham who could recognize talent when they saw it, and were in a position to help it progress.

Isaac seems to have been as lonely at school as he had been with his grandparents. He enjoyed the company of girls more than boys, and one of them later recollected that he was a 'sober, silent, thinking lad'. He did get into one famous fight with a bigger boy, rousing himself to such a passion that he gave the larger lad a thorough thrashing, and established a reputation which prevented him from being bullied. But his greatest delights were solitary study and manufacturing mechanical devices, including a working model of a windmill. He also caused one of the earliest recorded UFO scares by flying a kite at night with a paper lantern attached to it, thereby causing 'not a little discourse on market days, among the country people, when over their mugs of ale'. To his schoolboy companions, he must have seemed too clever by half, and was never liked.

At the end of 1659, when Newton was approaching the age of 17, he left the school to learn the management of the land that he would one day inherit from his mother. He was absolutely hopeless as a farmer, neglecting his work in order to read books and being fined on several occasions for allowing animals in his care to wander and damage other farmers' crops. One anecdote records that on the way from Grantham to Woolsthorpe there is a steep hill, where it was usual to dismount and lead your horse to the top before remounting; the story is that Isaac arrived home on foot with a book in one hand and the bridle in the other, his horse trailing along behind him, having forgotten to remount.

All the while Isaac was making a mess of farming, Hannah's brother William had been urging her to let the boy go to university, and the schoolmaster in Grantham, Henry Stokes, was so eager not to lose his star pupil that he offered to take him into his own home, without being paid for lodgings, if she would let him return to Grantham. Eventually, she agreed. In the autumn of 1660 (the year of the restoration of the monarchy in England), Isaac went back to school, specifically to prepare for entrance to Cambridge. In June 1661, undoubtedly with the benefit of the advice and influence of Humphrey Babington, he presented himself at Trinity College, and on 8 July he was formally admitted to the university. At 18, he was older than most new undergraduates; wealthy young gentlemen were still at that time often admitted at the tender age of 14, accompanied by a servant to look after them.

Some idea of the (lack of) enthusiasm with which Newton's mother viewed the enterprise can be gleaned from the fact that, far from having his own servant, he entered the college at the lowest level, as a subsizar, and paid for his keep by himself acting as a servant. Newton's own allowance was £10 a year, from a mother whose income by now was about £700 a year. At its worst, being a subsizar could be extremely unpleasant, and include the duty of emptying the chamber pots of the more affluent students; but the evidence suggests that Newton did not have to sink that low, and that he was technically the servant of Humphrey Babington (who was seldom in residence in

Cambridge), and so had few menial duties to perform. Nevertheless, Newton can hardly have enjoyed this status.

From what information we have, Newton seems to have spent a miserable first year and a half in Cambridge, lost in solitary study and lacking any real friends. Things only improved when, early in 1663, he met Nicholas Wickins. Newton and Wickins were both unhappy with their room-mates, and agreed to room together; the arrangement suited both of them so much that it lasted for twenty years. There is persuasive circumstantial evidence that Newton was a homosexual, either practising or latent, which may have been a factor in a personal crisis that we discuss later. Certainly the only close relationships he had were with men, but there were very few even of those. What little we know of Newton's way of life in those years comes from Wickins, via his son, John; none of Newton's other contemporaries as Cambridge undergraduates left any recollection of him, confirming that he was still a quiet thinker who kept himself to himself.

In fact, being a solitary thinker who loved books was the best way to learn in Cambridge in those days. Only a third of the students who entered left with a degree, even though about the only requirement for graduating was to stay there for four years. Although it had been one of the best universities when it was founded in the thirteenth century, by the second half of the seventeenth century Cambridge was by European standards a backwater where Aristotle was still taught by rote, and the only relevant training it gave was for a career in the church or (barely) medicine. There had been nothing there to compare with the circle of scientists that Wilkins drew to Oxford; the first scientific professorship in Cambridge, the Lucasian Chair of Mathematics, was established by an endowment from Henry Lucas in 1663, and awarded to Isaac Barrow, previously a professor of Greek. And, more than thirty years after the publication of Galileo's *Dialogue*, the Sun-centred Copernican cosmology was still disapproved of in Cambridge, in spite of the usual English love of anything the Pope didn't like.

Whatever Newton learned, he learned on his own, studying the writings of Galileo and René Descartes (1596–1650), among others. He was awarded a scholarship in April 1664, presumably because he was a good student (and very possibly partly due to Babington's influence), and automatically received his BA in January 1665. The scholarship provided him with the promise of a continued place at Cambridge until 1668, when he would become an MA; during that time, he could study whatever he liked (or, indeed, nothing at all).

Around that time, between receiving the scholarship and graduating, Newton's love of experimentation led him to use himself as a guinea pig in a series of alarming tests. First, he stared at the Sun with one eye for as long as he could bear it, in order to study the after images produced when he looked away; he nearly blinded himself (*never* try this yourself!) and only recovered his sight after being shut up in the dark for several days. Nothing daunted, a little later he poked a bodkin (a kind of blunt needle) into his eye, between the ball of the eye and the bone of the eye socket, and used it to press on the eyeball to observe the way the distortion affected his sight.

It is possible that Newton's interest in mathematics (and through that, science) was stimulated by the first series of lectures given by Barrow as Lucasian Professor, in the spring of 1664. It is certain that he became so absorbed in his studies that, Wickins tells us, he would forget to eat and would stay up all night at his books, becoming (while still in his twenties) the epitome of the absent-minded professor. He was careless about his dress, and there is no evidence that he ever took a bath (not that that was so unusual in seventeenth-century England).

By the time he graduated, or soon after, Newton was already well on the way to developing the calculus, and had (as we have seen!) begun his study of light. Then came one of the most famous episodes in his life.

The plague arrived in Cambridge in 1665, and led to the temporary closure of the university. Some students moved out to nearby villages with their tutors, and tried to carry on a semblance of normal life. But Newton had his BA, and worked alone in any case. So in the summer of 1665 he returned to Lincolnshire, where he stayed until March 1666. The spread of the disease had stopped during the winter months, and it seemed safe to return to Cambridge; but when the warm weather came back so did the plague, and in June Newton left for Lincolnshire again, returning to Cambridge permanently in April 1667, when the plague had run its course.

The period in Lincolnshire is often referred to as Newton's 'Annus mirabilis', even though it lasted for the best part of two years; popular accounts also depict him at Woolsthorpe for most of this time, although in fact, as well as the abortive return to Cambridge in 1666, he spent some (perhaps much) of the time at Humphrey Babington's rectory in Boothby Pagnell. No matter; the important point is that in Newton's own words, written half a century later, 'in those days [1665–6] I was in the prime of my age for invention & minded Mathematicks & Philosophy more than at any time since'. By the end of 1666, he was just 24 years old.

Although the mathematical achievements were breathtaking for a young man working entirely on his own, we shall skip over them; everybody knows that differential calculus is important, but only mathematicians care about how Newton came to invent it. What matters is that it provides a convenient way to deal mathematically with things that vary, such as the changing position of a planet in its orbit as time passes. But what everybody knows about this period of Newton's life is the story of the apple. This story, recounted by Newton many years later and intended both to show him in a good light and to claim precedence over Hooke, should be taken with a pinch of salt, but may be half true.

If there was an apple, it certainly didn't fall on Newton's head, and the completion of the link between falling apples, the Moon, and the inverse-square law of gravity only came much later. But the importance of the story is that during the plague years, in Lincolnshire, Newton realized that the influence pulling an apple downward from a tree and the influence holding the Moon in its orbit around the Earth might be one and the same.[3] He realized that the Earth's influence extended out into the Universe. What Newton puzzled over first was the question of why the spinning Earth didn't break up and fly apart in pieces because of centrifugal force – in the mid-1660s, this was still an argument used by some Aristotelians to deny that the Earth rotated at all. Newton's response was to calculate the outward force at the surface of the Earth due to its rotation, and to compare this with the measured force of gravity, showing that gravity is nearly three hundred times stronger at the surface of the Earth, so there is

3. It's also worth remembering that Hooke's *Micrographia*, which Newton read in 1665, suggested that the Moon possessed a 'principle of gravitation' like that of the Earth.

no puzzle about how the spinning planet stays intact. Once again, experiment and observation triumphed over mere philosophizing.

It was only after he returned to Cambridge (but some time before 1670) that Newton studied what he called the 'endeavour of the Moon to recede from the centre of the Earth' by virtue of its motion in its orbit, another puzzle involving what Christiaan Huygens had recently dubbed the centrifugal force. Why did the Moon stay in its orbit and not fly off into space?

This is a deeper question than it seems to modern eyes. Only a generation before, Galileo had been the first person to understand inertia, and to realize that a moving object will keep moving unless some force acts to slow it down. But, as we have seen, even he thought that the natural tendency of a moving object is to move in circles, explaining why planets stayed in orbit around the Sun and the Moon stayed in orbit around the Earth. Descartes realized that the natural tendency is for a body to keep its state of motion unless it is interfered with from outside – as Newton put it in one of his early notebooks:

> Every thing doth naturally persevere in y^t^ state in w^ch^ it is unlesse it bee interrupted by some externall cause . . . [a] body once moved will always keepe y^e^ same celerity, quantity & determination of its motion.

But be careful how you interpret that; when Newton was an undergraduate, such a statement could be understood as supporting Galileo's ideas about circular motion. It was Hooke who first appreciated that the natural state of a body is to keep moving *in a straight line* unless acted upon by some external force, and Hooke who expressed this idea clearly prior to Newton's definitive later work. So Newton's generation was the first (and, as it turned out, only) one for which there was a real puzzle about why the Moon did not fly off into space.

At first, Newton thought in terms of a centrifugal force trying to fling the Moon away into space, and a balancing force pulling it towards the Earth. His early work on the subject included a calculation which showed that the downward force of gravity at the surface of the Earth is four thousand times greater than the force needed to balance the centrifugal force appropriate for the Moon in its orbit. The two forces would exactly balance, at the distance of the Moon from the Earth, if gravity fell off as an inverse-square law; but Newton did *not* explicitly write this down at the time. He did, though, note that 'the endeavours of receding from the Sun [of the planets] will be reciprocally as the squares of the distances from the Sun'.

In other words, a planet twice as far from the Sun 'endeavours to recede' one quarter as strongly, a planet three times as far feels one ninth of the force, and so on. The implication was that if a force proportional to one over the distance squared was trying to fling the planets out of their orbits, and yet they stayed in orbit, there must be an exactly equal and opposite force holding them in place – gravity. But Newton didn't spell that out in the late 1660s, and was clearly still far from clear in his own mind about the nature and role of gravity. As we have seen, he first came to the attention of scientists outside Cambridge through his study of light and colours.

Newton's investigation of colours began in Cambridge before the plague years, and was largely abandoned while he was in Lincolnshire. But once back at Trinity, he completed his investigations, using triangular prisms to break up light from the Sun into the rainbow pattern of the spectrum. Earlier investigators, including Robert Hooke, had projected the spectrum on to a screen a couple of feet away, seeing a white spot of light with coloured edges. Newton projected his spectrum on a wall 22 feet away from the prism, and saw a spectrum five times as long as it was wide, which could not be explained as a coloured image of the Sun. The received wisdom about colours at that time was that white represented pure unsullied light, and that colours were formed by adding in increasing amounts of black to the white. Newton showed that white light is not pure at all, but is made up of a mixture of all the colours of the rainbow (black is simply the absence of light), and developed a new theory of light and colour, based on the idea of light being a stream of tiny particles.

Newton's demonstration that white light is a mixture of all the colours of the rainbow is a particularly neat example of the experimental method at work in science. A triangular prism would spread white light out into a rainbow spectrum. Newton said this was because the white light contained all the colours, and the path of each colour was bent by a different amount as it passed through the prism. Others suggested that the beam of pure white light had become coloured because of the effect of the glass on the light as it passed through – as if the light had got dirty. If Newton's idea was correct, a second triangular prism, placed upside down next to the first one, ought to bend the different coloured rays of light back into a single beam of white light. If the alternative hypothesis were correct, the second prism ought to make the light even 'dirtier' and more colourful. The experiment showed that Newton was correct.

The most astonishing thing about all this – three major breakthroughs, in different areas of science, any one of which would have made Newton's reputation – is that at the time Newton hugged all his results to himself. He seems to have been genuinely uninterested in fame, and to have studied nature simply for his own interest, for the pleasure of finding things out. But the way he hid his own light under a bushel for so long is particularly odd because in 1667 he could have done with a little recognition.

That year, there was an election of Fellows at Trinity. This was not a common event – there had been no elections for the previous three years – and there were only nine vacancies. Election as a Fellow would give Newton a few more years' security. The first rung on the ladder was a minor Fellowship, which would automatically become a major Fellowship when he became an MA in 1668. Then, he would have a further seven years to do anything he liked (or nothing at all, which is what most Fellows did) before being required under the statutes to take holy orders (in the seventeenth century, Cambridge University was still very much a Church-oriented establishment). In spite of hiding his scientific light under a bushel, though, Newton did receive one of the coveted Fellowships. Why?

There is one clue as to why Newton should have succeeded in being elected as a minor Fellow in October 1667, when his work was completely unknown, and most such appointments depended crucially on influence and patronage. Earlier that year, Humphrey Babington had become one of the Senior Fellows of the College.

The oath that Newton swore on taking up his Fellowship included the promise that 'I will either set Theology as the object of my studies and will take holy orders when the time prescribed by these statutes arrives, or I will resign from the college'. This was storing up trouble, because Newton was an Arian (see page 20). Although he would not suffer the same fiery fate as Giordano Bruno if this ever became public knowledge, it would certainly render him ineligible for the priesthood within the Anglican Church, which was firmly rooted in the idea of the Holy Trinity; it also made his status as a Fellow of Trinity College somewhat ironic, even as it seemingly set a time limit on how long Newton could hold on to that status. Newton was quite prepared to keep quiet about his beliefs if that enabled him to hold on to his status at Trinity; but he was not prepared to swear an oath that he believed in the Holy Trinity, which he would have to do if he took holy orders.

But at least he was secure for the time being. Newton used that security to continue his solitary studies in mathematics and his investigations of light and colour, becoming (if anything) increasingly absent-minded and vague in everyday matters. He had only three friends: John Wickins, Humphrey Babington, and Isaac Barrow; and we know tantalizingly little about any of the relationships, although clearly Newton must have discussed mathematics with Barrow, who became aware of his ability.

There are many stories of how Newton was often so deep in thought that he would sit down for dinner at high table in the hall and forget to eat anything during the entire course of a meal, never took any exercise (begrudging the time away from his studies), and, on the rare occasions he had a visitor, might step into his study for a bottle of wine, forget the purpose of his mission, and sit down at his desk to work. But the other Fellows were well aware that they had a rare genius in their midst – when Newton walked in the Fellows' garden and drew diagrams in the gravel of the path with his stick, the other Fellows would carefully avoid treading on the drawings, which stayed visible for weeks.

Bizarrely to modern eyes, however, towards the end of the 1660s Newton developed a consuming passion for alchemy, which would absorb him in the study of what were even then thought of as disreputable texts and the carrying out of experiments with various noxious substances, off and on (mostly on) for years.[4] Over the next two decades, he devoted far more time and effort to these studies than he

4. When Newton's hair turned grey in the 1670s, when he was only in his thirties, he joked to Wickins that it was thanks to his experiments with quicksilver (mercury) from which 'he took so soon that Colour'.

ever did to science. We have no space to discuss those activities here. Although they may be important to an understanding of Newton the man, they are less important for an understanding of the revolution in science in the seventeenth century; but it is important to appreciate that Newton had largely stopped being what we would call a scientist by the time he became famous.

It was just at this time (in 1669) that the outside world of science showed the first sign of breaking in on Newton's isolation. The mathematician Nicolaus Mercator (not the cartographer, Gerardus Mercator) published a book which took some of the first steps down the mathematical road Newton had already trodden. John Collins, a government clerk in London who was a kind of maths groupie, sent a copy to Isaac Barrow, who knew enough about Newton's work to realize that his priority was threatened. Under pressure from Barrow, for the first time Newton suffered the anguish he was to experience whenever he was asked to publish his work. He prepared a paper, going far beyond Mercator but still only hinting at how far he had actually gone, and sent it to Collins. Then, he asked for it back, deciding not to publish. But Collins knew enough about maths to recognize a work of genius, and took a copy, which he kept and showed around, perhaps a little indiscreetly. Word began to spread in the trade that there was a new mathematician of note in Cambridge.

This flurry of activity, abortive though it seems, may have been important in giving Newton his next step up the academic ladder. In 1669, Barrow decided to resign his chair. His motives are unclear, but seem to have been a mixture of a genuine desire to devote more time to divinity, and an unholy ambition to do better for himself; it is certainly not true that he simply decided that Newton was a genius who deserved the job more. There is nothing about the seedy and self-serving atmosphere of Cambridge in those days, nor about the known character of Barrow, to suggest such altruism. Since Barrow soon became a Royal Chaplain and then, in 1673, Master of Trinity College, it was clearly a good career move.

Barrow's influence was amply sufficient to ensure that the man he wanted replaced him as Lucasian Professor, and the man he wanted was Isaac Newton, whom he now knew to be the ablest mathematician in Europe. The post was one of the most desirable in Cambridge. It brought with it an income of £100 a year (remember that as a Fellow Newton already had free board and lodging, and a little additional income), with no tutoring responsibilities and the sole requirement being to give one course of lectures a year (even this requirement was often abused). It was a secure position for life – and it was Newton's at the age of 26. There was one peculiarity about the bequest under which Henry Lucas had established the chair – it specifically barred any holder from accepting a position in the Church requiring residence outside Cambridge or 'the cure of souls'. This, as we shall see, was to prove a godsend to Newton.

Newton must have imagined that he could now return to a life of quiet anonymity. Alongside his passion for science, he now developed an equal zeal for theology, of a distinctly unorthodox kind. It isn't clear at what point he became an Arian. He may have found the idea attractive and turned to theology as a result, or he may have started studying theology with a view to taking holy orders, and then decided that there was something wrong with the established interpretation. Either way, his self-appointed task became seeking out evidence from the scriptures and other writings to support his conviction that Arianism was the true path, and that Trinitarians had taken a wrong turning way back in the early centuries after Christ. Again, this is not the place to go into details. But it is easy to imagine that Newton could have become a complete recluse, stuck in his Cambridge rut, who never published anything and whose scientific work only came to light, if at all, after his death. He was saved from that fate by his own inventiveness, and Barrow's interest in his work, which were soon to thrust him firmly into the spotlight.

It was Newton's work on optics and colour in the late 1660s that led him to think about telescopes and the problem that the lenses produced coloured fringes around the image of the object being viewed. From his experiments, Newton realized that there was no way that these fringes could be avoided using simple lenses (in fact, modern telescopes do solve the problem by using compound lenses made up of more than one kind of material, but this was way beyond the technology available to Newton). But he also realized that a reflecting telescope, in which the magnifying is done by a curved mirror, would not suffer from this problem.

Characteristically, in order to demonstrate the value of a reflecting telescope, Newton built one himself. We now know that he wasn't the first – Leonard Digges built one, probably on the same principles as Newton's telescope, before 1550. But Newton knew nothing of this, and produced his design independently.

The problem with a reflector is that in its simplest form the head of the observer would be in the mouth of the telescope, blocking out any incoming light. James Gregory's design, in 1663, used two curved mirrors, the second a small one in the mouth of the telescope to reflect light from the main mirror back through a hole in the middle of the main mirror. The Gregorian telescope design became widely used in astronomy, but only much later. In Newton's simpler design, a small flat mirror at an angle of 45° in the mouth of the instrument deflects light from the main mirror sideways, where the observer can see it comfortably without interfering with the operation of the instrument. Newton built his instrument, using a metal mirror he shaped and polished himself.

In fact, he may have built two, one around 1667 or 1668, of which no more is heard, and one in 1671 (possibly, it was one telescope that he put aside for a couple of years and then got out again). It was about 6 inches long, and allegedly enlarged as well as a refractor 10 feet long. Late in life, Newton was asked where he got it made, and replied that he had made it himself. His questioner then asked where he got the tools, and he laughed and said that he made them as well, for 'if I had staid for other people to make my tools & things for me, I had never made anything of it'.

Collins learned about the telescope before the end of the 1660s, and word spread both from him and from visitors to Cambridge who had seen the instrument. As we have seen, the Royal Society heard of it before the end of 1671, and asked for a demonstration of its powers. It was Isaac Barrow who took the telescope to London and showed it to them.

Ironically, although Newton's telescope was an immediate sensation, and made his name, it never proved possible in Newton's time to make a reflector big enough, and with an accurate enough mirror, for serious astronomical work. The immediate effect of his sudden acceptance as a leading scientist was that he felt secure enough to send his theory of light and colours to the Royal for publication, which led to the bitter row with Robert Hooke that we have already described.

Whatever the rights and wrongs of that dispute, Newton was furious that anybody should doubt his word and belittle his ideas, and equally angry at the time he had to waste defending his theory of light, when he could be doing new work. He threatened to resign from the Royal, and when persuaded to stay instructed the Secretary, Henry Oldenburg, not to forward any correspondence to him because 'I intend to be no further solicitous about matters of Philosophy'. He withdrew once again into his shell in Cambridge, where he began his serious study of theology, alongside his alchemical experiments. By then, he was more than halfway through the initial seven-year period of his major Fellowship, and coming up to the time when he would have to take holy orders. Looking for a loophole, before long Newton convinced himself that the whole foundation of established religion in England was based on a corruption of the original biblical texts, and specifically that the concept of the Holy Trinity, placing Jesus Christ on an equal footing with God, was a false one. To Newton, worshipping Christ as God would be idolatry, a mortal sin that would put his soul in peril. But without ordination, there would be no continuing membership of Cambridge University, and therefore no Lucasian Chair. It would be back to the farm in Lincolnshire.

All of this was vastly more important to Newton as 1675 approached than any quibbles about optics or mathematics. His more and more detailed researches into theology would continue obsessively for many years, but like the alchemy they have no real place in our story except to help to explain why he was so secretive in general. By the beginning of 1675, the situation looked so desperate that Newton wrote to Oldenburg requesting to be excused from paying his subscription to the Royal, because 'I am to part with my Fellowship, & as my incomes contract, I find it will be convenient that I contract my expenses'.

But he had one last fling of the dice to try. Isaac Barrow was by now Master of Trinity, and without explaining his true reasons for the request, Newton obtained his permission to petition the King for a dispensation, not for himself alone as a special case, but for *all*

Lucasian professors, from the requirement of ordination. His argument was that even ordination would be against the spirit of the Lucas bequest, with its specific requirement that a holder of the post should not be active in the Church. The King (still Charles II) had the power to dispense with any statute of the university, for any reason he liked (or none at all). He was interested in science, patron of the Royal Society, and granted the dispensation in perpetuity, 'to give all just encouragement to learned men who are & shall be elected to the said Professorship'. Newton was safe. After this crisis, plus the unpleasantness with Hooke, it is small wonder that he lay low in Cambridge, deep in his theological and alchemical studies, and generally let the world of science get on with things in its own way – a decision no doubt reinforced when Oldenburg died and Hooke was elected as his successor as Secretary of the Royal Society in 1677.

Just in case you think Newton was being obsessively cautious in going to such lengths to conceal his Arianism, it's worth noting that his eventual successor as Lucasian Professor, William Whiston, was dismissed from his post in 1710 for publicly discussing what he saw to be errors in the Anglican faith.

It's against this background that the famous debate with Hooke took place, culminating in the 'shoulders of Giants' letter of 1676; Newton's preoccupation with alchemy and theology and concern for his own future at Trinity offering some explanation for why he was so irritated at being dragged into a public debate about priority in science. But while Newton was retreating into his shell, Hooke was becoming more active in science again after the years he had spent involved in the rebuilding of London after the Great Fire. In 1674, he published his latest thoughts on orbital motion, in which he rejected the idea that the planets might be held in their orbits around the Sun, and the Moon might be held in its orbit around the Earth, by a balance of forces, one tending to push it outward and the other tending to pull it inward. Previously, Newton, Huygens, and others had thought in terms of objects like the Moon and planets having a 'tendency to recede' from the centre of the object they were orbiting. As we have seen, Hooke realized, and spelled out, that orbital motion could be explained by the natural tendency of any object to move in a straight line unless it is interfered with, plus a *single* force acting towards the centre of the attracting body – the Sun or the Earth. He also had another profound insight. Previously, following the ideas of Descartes, everyone (including Newton) had thought in terms of the Moon and planets being carried along by some kind of swirling, invisible fluid (like chips of wood in a whirlpool); it was Hooke who introduced the idea of what came to be called 'action at a distance' – the idea that gravity could reach out across *empty* space to act on the Moon and planets.

All of these ideas, a 'System of the World' based on 'three suppositions', were described lucidly in one of Hooke's Cutlerian lectures, in 1674:

> First, that all Coelestial Bodies whatsoever, have an attraction or gravitating power towards their own Centers, whereby they attract not only their own parts, and keep them from flying from them . . . but that they do also attract all the other Coelestial Bodies that are within the sphere of their activity . . . The second supposition is this, That all bodies whatsoever that are put into a direct and simple motion, will so continue to move forward in a straight line, till they are by some other effectual powers deflected and bent into a Motion, describing a Circle, Ellipsis, or some other more compounded Curve Line. The third supposition is, That these attractive powers are so much more powerful in operating, by how much nearer the body wrought upon is to their own Centers.

At this stage, Hooke is still thinking of gravity as a linear force, not as an inverse-square; but he was only five years away from getting that right, as well.

Newton seems not to have been aware of these ideas at the time, having deliberately turned his back on science. He rarely left Cambridge in the years that followed, the notable exception being in 1679, when his mother became ill, and he went to be with her in Woolsthorpe while she died. As both her heir and executor, he had to spend several months putting the affairs of the estate in order. In the same year, Hooke, as Secretary of the Royal, wrote to Newton asking for his opinion on the ideas about orbits and gravity that Hooke had published five years earlier, in particular the idea of 'compounding the celestial motions of the planets' in terms of 'a direct motion by the tangent and an attractive motion towards the central body'. Newton's reply was a polite brush-off, mentioning his recent preoccupation with affairs in Woolsthorpe:

> [I] had no time to entertain philosophical meditations, or so much as to study or mind anything else but country affairs. And before that, I had for some years past been endeavouring to bend my self from Philosophy to other studies in so much y^t I have long grutched [grudged] the time spent in y^t study unless it be perhaps at idle hours sometimes for a diversion . . . I hope it will not be interpreted out of any unkindness to you or y^e R. Society that I am backward in engaging myself in these matters.

Newton claimed never to have heard of Hooke's ideas about planetary motion before this date, 'though these no doubt are well known to the philosophical world'. Nevertheless, he was intrigued, and, almost in spite of himself, got drawn into thinking about gravity and orbits again. It is noteworthy that in all Newton's subsequent work on the subject the idea of action at a distance is used, without comment or credit, together with the assumption that an orbit is a straight line that has been bent by gravity. But Newton was able to see further than Hooke because he had the ability to take physical ideas and turn them into mathematical problems by setting up systems of equations that applied to the physical systems of interest, and could be solved mathematically. This was a key step in the scientific revolution.

The immediate effect of this prompting by Hooke was that in the same letter Newton discussed a way to measure the rotation of the Earth. He ended the letter with the protestation:

> But yet my affection for Philosophy being worn out, so that I am almost as little concerned about it as one tradesman uses to be about another man's trade or a country man about learning, I must acknowledge my self avers from spending that time in writing about it W^ch I think I can spend otherwise more to my own content.

The trouble was, Newton's description of his new idea included a careless slip of the pen, which produced a reply from Hooke correcting the error and drew Newton into more correspondence, in which Hooke made the telling point that although Newton's calculation was based on a force of attraction with 'an aequall power at all Distances from the center', Hooke's own 'supposition is that the Attraction is always in a duplicate proportion to the Distance from the Center Reciprocall'. That is, an inverse-square law. It was Hooke who introduced the idea of an inverse-square law to Newton.

Newton never replied to this letter. But all the evidence suggests that it was this comment by Hooke, made in 1680, that inspired Newton to *prove* that an inverse-square law of gravity was the only law that could explain the orbits of the planets around the Sun and the Moon around the Earth. Newton's ability to prove this, when Hooke could only surmise it, demonstrates that Newton was a greater mathematician than Hooke (though not necessarily a better scientist). But it would be a further four years before another intruder from the world outside Cambridge prompted Newton to make this proof public.

The outside world of science also intruded on Newton during the winter of 1680–81, though, when a new comet was seen. The Astronomer Royal, John Flamsteed, realized that what was observed was one comet first approaching the Sun, being lost in the Sun's glare, and then retreating from the Sun. Hooke, as we have seen, had already suggested this in connection with the comets he had studied with Wren in 1664–5, but the idea was still new, and Flamsteed's espousal of it carried more weight than Hooke's original suggestion. As so often, a major insight provided by Hooke often gets credited in the history books to a later scientist. Whatever, previously it had been thought that such events were caused by two separate comets. But Flamsteed imagined some kind of magnetic repulsion, in which the comet was turned around before reaching the sun. Newton, like Hooke but perhaps independently, realized that the comet must have gone round the far side of the Sun in a curving path more like an elongated planetary orbit. Almost in spite of himself, Newton's attention had been drawn back to thinking about gravity and orbits in the early 1680s. The interest was fanned further by the appearance of another comet (now called Halley's Comet) in 1682.

Then came another disturbance in his private life. Newton's long-time friend, room-mate, and scribe John Wickins decided to resign his Fellowship in 1683. He left Cambridge, became Rector of Stoke Edith, in Herefordshire, and married. The limited correspondence between the two suggests some bitterness about the breakup of the old friendship, at least on Newton's part. On the practical side, Wickins was quickly replaced by a young man from Newton's old school in Grantham, his namesake Humphrey Newton. Neither Humphrey nor Isaac claimed that they were related, but there may have been a distant connection. Humphrey stayed with Newton for five years as a kind of amanuensis, copying out his works and probably acting partly as a servant; it is from him that some of the more colourful accounts of Newton's life in Cambridge have come. Among other things, it was Humphrey Newton who made the fair copy of Isaac Newton's greatest work, the copy from which the *Principia* was printed.

The story of how Newton came to write his masterpiece begins with a meeting of the Royal Society in January 1684.[5] After the meeting, Edmond Halley, Christopher Wren, and Robert Hooke, keeping themselves warm in a nearby coffee house, had a discussion about planetary orbits. They were by now convinced that the orbits of the planets around the Sun, in accordance with the empirical laws of planetary motion discovered by Johannes Kepler early in the seventeenth century, obeyed an inverse-square law; but they could not prove that an inverse-square law of gravity *must* produce elliptical orbits, nor that this was the only law which produced elliptical orbits. For all they knew, it could just be a coincidence; perhaps one day a new planet would be discovered in a different kind of orbit (indeed, it still seemed possible at that time that comets were heavenly bodies in a different kind of orbit). Halley and Wren acknowledged that the task of proving that Kepler's laws followed inevitably from an inverse-square law of gravity was beyond them. Hooke, in a typical show of chutzpah, claimed that, if he put his mind to it, he could prove that an inverse-square law strictly required elliptical orbits. Wren promised to give either Hooke or Halley a book to the value of £2 if one of them could come up with the proof within two months. Halley soon admitted defeat; Hooke never actually admitted defeat, but never won the prize either. But the friendly rivalry between the two of them and Wren kept the puzzle at the forefront of their discussions over the next few months.

5. The winter of 1683–4 was, incidentally, one of the coldest ever recorded in England, when the Thames froze from bank to bank above London Bridge, and a tented 'Frost Fair' was held on its icy surface.

In the summer of 1684, Halley had family business in Peterborough (probably in connection with the death of his father; see the next chapter). Taking advantage of being in the area, and still baffled by the puzzle of planetary orbits, in August he went to Cambridge to see the man regarded as the best mathematician in Europe (even if he had been keeping quiet lately), and the one who could surely be relied on to solve the puzzle if anyone could. Although Newton was known to be difficult to get on with, Halley had already corresponded with him about comets, and the two may have met in 1682, so to some extent the ice had been broken. What happened at their meeting has been described by the mathematician Abraham de Moivre, a French Huguenot refugee and acquaintance of Newton, reporting the events as Newton described them to him. This account, written much later (Halley wasn't a 'Dr' and Newton wasn't a 'Sir' in 1684!), should not be taken as the literal truth (Newton always reported such events to show himself in the best light), but is the nearest we will ever get to eavesdropping on one of the most important conversations in the history of science:

> In 1684 Dr Halley came to visit him in Cambridge, after they had been some time together the Dr asked him what he thought the Curve would be that would be described by the Planets supposing the force of attraction towards the Sun to be reciprocal to the square of their distance from it. Sr Isaac replied immediately it would be an Ellipsis, the Dr struck with joy & amazement asked him how he knew it, why saith he, I have calculated it, whereupon Dr Halley asked him for the calculation without any further delay, Sr Isaac looked among his papers but could not find it, but he promised to renew it, & then send it to him.

Of course, the paper was not lost. Newton was playing for time, reluctant, as ever, to let the world know of his discoveries and endure the time-wasting public attention that would result. But this time, perhaps still motivated by the desire to put one over on Hooke, he succumbed. When he did dig out his calculations, he found a small error, and sat down to work them out again. As always, once he became embroiled in a problem he could not stop. In November 1684, Halley received a nine-page paper not only proving the relationship of the inverse-square law of gravity to planetary orbits, but hinting that this was just the tip of the iceberg of a larger body of work.

The paper, *On the Motion of Bodies in an Orbit*, was a dramatic step forward, which Halley presented to the Royal Society on 10 December. Encouraged by Halley (with the enthusiastic support of the Royal), and gripped again by his desire to get everything sorted out in his own mind, Newton even abandoned his alchemy for the time being, and threw himself into writing what became his great book, *Philosophiae Naturalis Principia Mathematica*. Halley nurse-maided the whole enterprise, cajoling Newton when he became reluctant to finish the work, and soothing him when (inevitably) Hooke, who saw the draft of the book in his capacity as Secretary of the Royal, complained (justifiably but perhaps tactlessly) that Newton had given him insufficient credit. Halley explained the situation in a letter to Newton:

> Mr Hook has some pretensions upon the invention of ye rule of the decrease of Gravity, being reciprocally as the squares of the distances from the Center. He sais you had the notion from him, though he owns the Demonstration of the Curves generated therby to be wholly your own; how much of this is so, you know best, as likewise what you have to do in this matter, only Mr Hook seems to expect you should make some mention of him, in the preface, which, it is possible, you may see reason to praefix.

This seems entirely reasonable, not least since we know that Newton did get both the inverse-square law and the idea that an orbit is a straight line bent by gravity from Hooke. But his reaction was far from reasonable. After initially threatening to suppress the publication of his work entirely, Newton went through the manuscript of the third volume at a late stage, savagely removing almost every reference to Hooke; the amended manuscript survives, providing clear evidence of Newton's attempt to write Hooke out of the story. In one of the few places where Hooke's name remains, where it appeared originally as a reference to 'Clarissimus [most distinguished] Hookius' the 'Clarissimus' has been deleted.

Not all of Newton's contemporaries, or later historians, were fooled. Aubrey (admittedly a friend of Hooke) describes the idea that 'gravitation was reciprocall to the square of the distance' as 'the whole coelastiall theory, concerning which Mr Newton haz made a demonstration, not at all owning he receiv'd the first Intimation of it from Mr Hooke'.[6] And a modern historian, Michael Hunter, describes Hooke more objectively as 'the unfortunate victim of the virtual deification of Newton since the eighteenth century'.[7]

While the argument simmered in the background, Halley placated Newton as much as possible, read the proofs of the book, and even paid for the publication of the *Principia* when the Royal Society, in desperate financial straits, found that it had insufficient funds for the task (happily, Halley actually made a small profit on the deal). The great work was eventually published in July 1687, completing the scientific revolution that had been begun by Gilbert and Galileo.

It's worth recapitulating the key insights that Newton spelled out in his book, while remembering that he was not the sole progenitor of these ideas; what mattered was that he pulled everything together and set it in a logical framework, indeed setting out the 'mathematical principles of natural philosophy'. As well as his law of gravitation – a *universal* law of gravitation, which applied to all material objects, anywhere in the Universe, at any time – Newton provided in the book the three laws of motion which formed the bedrock on which physics could be built. The first law states that every object continues to be at rest, or to move in a straight line at a constant speed, unless it is acted upon by an external force. The second law states that the acceleration of an object (the rate at which the velocity changes, which means either its speed, or the direction it is moving in, or both) is proportional to the force acting on it. And the third law states that whenever a force is applied to a solid object, it responds with an equal and opposite force – if you press your finger against the wall, the wall resists your push.

There is another fundamental piece of physics in the *Principia*, which we have already mentioned but is not always given the attention it deserves. The gravitational force considered by Newton, following Hooke's proposal, was an action at a distance. He did not require the

6. Hooke's own response was delicious, if a little biased. In one of his lectures he referred in passing to 'those proprietys of Gravity which I myself first Discovered and shewed to this Society many years since, which of late Mr Newton has done me the favour to print and Publish as his own Inventions'.

7. See Bennett, Cooper, Hunter, and Jardine, *London's Leonardo*.

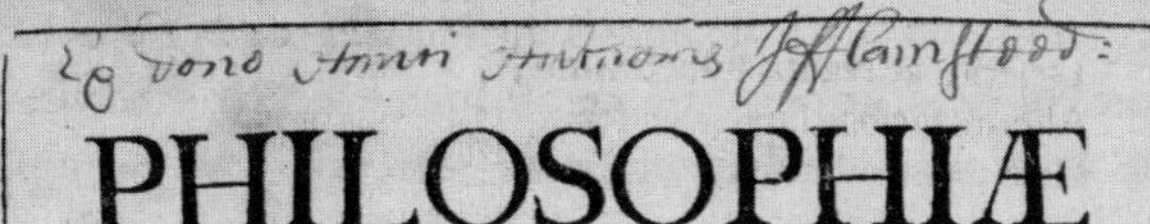

PHILOSOPHIÆ
NATURALIS
PRINCIPIA
MATHEMATICA.

Autore *JS. NEWTON*, *Trin. Coll. Cantab. Soc.* Matheseos Professore *Lucasiano*, & Societatis Regalis Sodali.

IMPRIMATUR.
S. PEPYS, *Reg. Soc.* PRÆSES.
*Julii* 5. 1686.

Soc. Reg. Lond.

*LONDINI*,

Jussu *Societatis Regiæ* ac Typis *Josephi Streater*. Prostat apud plures Bibliopolas. *Anno* MDCLXXXVII.

*Title page of Isaac Newton's* Principia

existence of any medium between the Sun and the planets to transmit the force of gravity – it was not as if the planets were tied to the Sun by elastic, or held in place by mysterious swirling vortices of some invisible fluid. We should emphasize that this was a dramatic and revolutionary suggestion at the time, because it soon became so deeply embedded as part of physics that modern readers don't think it odd at all. And we might remind you that Newton got the idea from Hooke.

The *Principia* made such a big impact because it finally established, for those with eyes to see, that the world ran on essentially mechanical principles, principles that could be understood by ordinary human beings, and that it was not governed by magic or the whims of capricious gods. And on many occasions Newton spelled out the scientific credo. Remember the letter to Gaston Pardies:

> The best and safest method of philosophizing seems to be, first to inquire diligently into the properties of things, and to establish those properties by experiences [experiments] and then to proceed more slowly to hypotheses for the explanation of them. For hypotheses should be employed only in explaining the properties of things, but not assumed in determining them; unless so far as they may furnish experiments.

As we pointed out earlier, this was no more than Gilbert had said almost a century before, and it may have been a conscious echo of Baconian sentiments. But where Gilbert, at the end of the sixteenth century, had been little more than a lone voice crying in the wilderness, Newton, at the end of the seventeenth century, was expressing the mainstream view of natural philosophers, endorsed by the Royal Society. Although scientists on the mainland of Europe, including Christiaan Huygens, were slow to appreciate the full importance of Newton's work, 'Newtonian' physics became the accepted template for the way of doing science in Britain as soon as thinking people had had time to digest the *Principia*.

Without doubt, Newton had now arrived. He was hailed as a genius, and could never again retreat into isolation. But he had also, in his 45th year, virtually stopped being a scientist. The *Principia*, great though it was, was a masterly summation of ideas (not all of them original to Newton) developed over the past twenty years. From now on, Isaac Newton would be a public figure, but although he would continue to impress the world with his scientific publications, they would almost entirely be old material that he had hugged to himself for years, waiting until the time was ripe for publication.

Even before writing the *Principia*, Newton had begun to be more active in college life at Trinity, as one of the more senior Fellows. By March 1687, the masterwork was largely off his hands, some portions already with the printers, one volume being diligently copied out by Humphrey Newton. And Newton had time to make his name in Cambridge in another way, as a principal defender of the rights of the university against the new Catholic King, James II, who had succeeded his brother in 1685.

There was still a strong anti-Catholic feeling in England, but by 1687 James had been on the throne long enough to begin to feel confident, and to try to throw his weight around. On 9 February, the University of Cambridge was ordered by the King to admit a Benedictine monk, Alban Francis, to the degree of Master of Arts, without requiring him to take any examinations or to swear any oaths. It was quite the done thing to confer degrees in this way (equivalent to a modern honorary degree) on someone like a foreign dignitary, and this had been done before even for Catholics. But the university was well aware that Francis intended to exercise his rights as an MA to participate in the activities of the university. One Benedictine would pose no problem; but surely there would soon be others, voting the Catholic ticket in the university's affairs.

Although the Cambridge academics feared the prospect, they were largely a bunch of feeble time-servers who lacked the guts to stand up to the King. Astonishingly, in view of the way he had previously avoided any kind of publicity or fuss (but perhaps understandably in view of his Arian beliefs), Newton became one of the most outspoken opponents of the proposal, and was largely instrumental in making the university refuse to accept Francis or award him an MA. This really did take some guts – as a ringleader in the opposition to the King, Newton was among nine Cambridge Fellows who had to appear before the notorious Lord Chancellor, 'Bloody' Judge Jefferies, who had had 300 people hanged for rebellion the previous year. Remember that this was before the publication of the *Principia*, when Newton was still largely unknown outside academic circles; at that time, he had no widespread fame to protect him from the potential consequences of Jefferies' wrath.

Newton achieved great eminence in Cambridge for this stand. When King James was removed at the end of 1688 and was replaced by the Protestant William of Orange (grandson of Charles I) and his wife Mary (daughter of James II) in 1689 (the 'Glorious Revolution'), Newton was one of two representatives from the University of Cambridge elected to the Convention Parliament, which legalized the position of William and Mary, and re-established the Anglican Church. The Parliament was dissolved after a year and a month; Newton had played no active part, but kept his mouth shut and voted the party line. He did, though, suffer one great disappointment.

One of the Acts passed by the Parliament increased the level of legal tolerance given to religious dissenters. Newton must have hoped that this would enable him to come out as an Arian. But the Act specifically excluded two classes of people from toleration – Catholics, and 'any person that shall deny in his Preaching or Writeing the Doctrine of the Blessed Trinity'.

Hiding his disappointment, Newton returned to Cambridge, and did not offer himself for re-election to the next Parliament. But while in London he had renewed and cultivated an old friendship, with Charles Montague (the grandson of the Earl of Manchester), who had been a student at Cambridge (indeed, at Trinity, which he entered in 1679) and became friendly with Newton then. Montague became a powerful political influence during the reign of William and Mary, and this soon totally changed Newton's life.

Not that it wasn't already going through changes. Also in 1689, Newton became firm friends with a young Swiss mathematician, Nicholas Fatio de Duillier (usually referred to as Fatio), who became something of a confidant of Newton's, and passed on news about Newton and his work to a wider audience of acquaintances, including Christiaan Huygens. Fatio had arrived in England in 1687, the year he was 23, and been elected as a Fellow of the Royal the same year. He first met Newton at a meeting of the Society in June 1689 – the same occasion on which Newton first met Christiaan Huygens. William of Orange had just become King of England, so the Dutchman Huygens was a particularly honoured guest at the Royal; Fatio, who had met Huygens on his travels in Europe, acted as his guide on his visit to London. The friendship between Newton and Fatio took root immediately, and remained very close for four years – so close, indeed, that it led to Fatio being asked to live with Newton in Cambridge and work with him on a second, bigger and better edition of the *Principia*. But nothing came of the suggestion, and by 1693 the relationship was petering out, at least on Fatio's side. In the first half of that year their correspondence reveals a growing tension in the relationship, and then Fatio simply disappears from Newton's life, with no surviving evidence to tell us about the final break.

But Newton had much else to preoccupy him in the early 1690s as he began to come out of his Cambridge shell. About the same time that he got to know Fatio, he also met, and struck up a lasting friendship with, the philosopher John Locke. This friendship was cemented by their common anti-Trinitarian views, and a shared interest in alchemy. Another friend, whom Newton had already met through the Royal Society, was, of course, Samuel Pepys.

With his horizons broadened by his experiences while serving as a Member of Parliament in London, and by friends such as these, Newton found returning to Cambridge life in the early 1690s frustrating. He threw himself into his alchemy, and also began gathering up the threads of all his more public work into a planned great volume summing up his contributions to science; but he also began sounding out opportunities to leave the university altogether and find some other employment. By 1693, his frustration at being unable to find such a post, years of overwork, the breakup of the relationship with Fatio, and the strain of keeping his Arianism concealed told at last, and he suffered a major nervous breakdown (it can't have helped that by then the tedious dispute with Wilhelm Leibnitz about who had discovered calculus first was rumbling along; this row is of no interest today, except for the way in which it may have disturbed Newton's peace of mind).

Few details of the illness survive, except for two bizarre letters, one to Locke and one to Pepys, apologizing for making wild accusations that his friends were conspiring against him, and trying to get him embroiled with women (Locke and Pepys were both known to enjoy female companionship). The letter to Locke, written in a shaky hand on 16 September 1693, is particularly bizarre:

Being of opinion that you have endeavoured to embroil me w$^{th}$ woemen & by other means I was so much affected with it as that when one told me you were sickly & would not live I answered twere better if you were dead. I desire you to forgive me this uncharitableness. For I am now satisfied that what you have done is just & I beg your pardon for having hard thoughts of you for it & for representing that you struck at y$^{e}$ root of mortality in a principle you laid down in your book of Ideas & designed to pursue in another book & that I took you for a Hobbist.[8] I beg your pardon also for saying or thinking that there was a designe to sell me an office, or to embroile me. I am

your most humble & most
unfortunate Servant.
Is. Newton

There have been suggestions that the mental illness was caused by Newton's close proximity for many years to noxious substances during his alchemical experiments, but this seems unlikely, because he soon made what seems to have been a complete recovery, which doesn't happen with, say, mercury poisoning. There is also some evidence that Newton had suffered similar, but lesser (or better concealed?) bouts of such disturbance before, notably in the months just after Wickins left Cambridge. This parallels the way the depression in 1693 followed the breakup of the friendship with Fatio.

Whatever the causes of the illness, Newton's friends stood by him and helped his recovery. In 1694, he took up an old interest in the problem of the exact orbit of the Moon around the Earth (since the Moon is influenced by the gravity of the Sun as well as the Earth, this is an example of the 'three body problem' in gravitation, which, it is now known, cannot be solved precisely; no wonder Newton said that his head 'never ached but with his studies on the moon'). But by the standards of his great days, this was mere tinkering around the edges of gravitational theory. It must have come as an immense relief when in 1696 Charles Montague, now Chancellor of the Exchequer and well aware of Newton's desire to leave Cambridge, offered him the post of Warden of the Royal Mint. He accepted with alacrity – the appointment was confirmed on 19 March, and before the end of April

8. That is, a follower of the materialist philosophy of Thomas Hobbes.

Newton had moved to London, lock, stock, and barrel. He would return to Cambridge for only a few days in his entire remaining life – although, having now learned what was expected of a Cambridge Fellow, he retained all his Cambridge posts and the income from them until 1701.

Montague's letter informing Newton of his appointment mentioned that the post was 'worth five or six hundred pounds [a year], and has not too much bus'nesse to require more attendance than you may spare'. It was, in other words, potentially a sinecure. But Newton was never one to shirk a job (except farming), and threw himself into his work. The Warden was actually number two at the Mint, which was technically run by the Master; but the Master, Thomas Neale, was quite happy to leave all the work to Newton. And work there was. Because of forgery and clipping of silver coins, the currency was debased and there was a serious monetary crisis. Parliament had just passed an Act authorizing a complete recoinage, and it fell to Newton to see this task through. The new technology which made the project worthwhile was a secret process which put a milled edge around the new coins, so that clipping them could easily be detected. Newton improved the efficiency of the Mint, cleaned out corruption, and completed the task by the summer of 1698. Montague told anyone who would listen that the job would never have been completed if it were not for Newton.

One of Newton's less palatable duties was to prosecute counterfeiters – when a successful prosecution usually meant a hanging. At first, he tried to get out of what he called a 'vexatious & dangerous' task, but when ordered by the Treasury to get on with it, he did so with his customary efficiency and a certain cold-blooded ruthlessness slightly chilling to modern eyes, but quite normal for someone in his position in the 1690s; he even became a magistrate, so that in effect he was both prosecutor and judge in the cases of counterfeiting that were uncovered. At the end of 1699, Thomas Neale died, and even though Montague was no longer in power Newton was quickly appointed as his successor – a unique promotion in the long history of the Royal Mint, which stands as a recognition of how well he had carried out his work. He held the post of Master until he died, although in his later years the real work was deputed to his eventual successor. The appointment was confirmed on 3 February 1700, and in 1701 Newton finally resigned from his Cambridge posts, having made £3,500 that year from his post at the Mint (on a complicated profit-sharing basis which we need not detail here, but all completely legal and above board, although this was an unusually good year). To put that into perspective, a calculation made in 1696 estimated the income of a nobleman as £3,200 a year, that of a baronet as £880, and of a country gentleman (the highest Newton could have aspired to if his mother's wish for him to take over the farm had come true) £280.[9]

With everything running smoothly at the Mint and the recoinage long since completed, in the 1700s Newton became active in other areas. Probably at the behest of Lord Halifax (as Charles Montague had now become), he stood for Parliament again in 1701, and was again elected to represent Cambridge, although once again his activity in Parliament was limited to voting as he was told to vote. In May 1702, William III died, and Parliament was dissolved. Mary had died in 1694, and William was succeeded by Queen Anne, the second daughter of James II. Anne was heavily influenced by Halifax, and it was during the election campaign of 1705 that she knighted both Newton and Halifax's brother, raising their profiles in a blatant attempt to win votes. It didn't help. The political tide was against Halifax and his party; Newton was not elected, and never stood for Parliament again. But it is worth recounting the story, since many people believe that Newton was knighted for his scientific achievements, some think that it was a reward for his work at the Mint, and few appreciate that it was a sordid party political ploy that didn't even achieve its intended aim. Newton was, though, the first scientist to be honoured in this way, if honoured is the right word.

By 1705, the year he turned 63, Newton was probably happy to be out of politics, since he had recently been elected to the post in which he would, though now in his sixties, make his last, and by no means least, mark on science – President of the Royal Society. The way had been cleared by the death of Newton's old adversary Robert Hooke, in March 1703 at the age of 67. As long as Hooke had been around, Newton had avoided having anything much to do with the Royal.

9. Quoted by Craig.

With the opposition out of the way, however, Newton was elected President on 30 November that year, and his second great book, *Opticks*, was published in 1704. The book had been completed in the mid-1690s, and Newton had quietly waited to publish it until Hooke had died and could not draw attention to the places where Newton had been influenced by Hooke's own work on light and colour.

If anything, the success of *Opticks* was greater than that of the *Principia*, because it was written in easily intelligible language and dealt with ideas, such as light and colour, that everyone could relate to. It was the crowning glory of Newton's scientific reputation, and helped to establish him as not just the most famous but also the most powerful scientist in the land. In it, Newton once again asserted the Baconian principles on which science ought to operate:

> Analysis consists in making Experiments and Observations, and in drawing general Conclusions from them by Induction, and admitting of no Objections against the Conclusions but such as are taken from Experiment, or other certain Truths.

This insistence on the importance of the scientific method, coupled with the shining example of the great discoveries that he had made using this method, was highly influential in getting the scientific method established, which was just as important for the future of science as the discoveries themselves.

Newton ruled the Royal with a rod of iron for more than twenty years, during which he missed only three of its meetings;[10] he was by no means universally popular among its members, although he always got a healthy majority of the votes for the Presidency, and held on to the post even after old age and infirmity reduced his contribution to the Society. Newton made the Royal genuinely scientific, and completed the task of turning a somewhat dilettante gentlemen's talking shop into a truly learned society.

This role has, though, sometimes been exaggerated, as part of the Newton myth. As we have seen, Newton was not the first person to attempt this professionalization of the Society, and he would certainly

10. His immediate predecessors had been mere figureheads, who between them had managed to attend only three meetings in the past eight years.

not be the last. But he did make strenuous efforts to keep the finances in order by dunning the arrears of subscriptions out of offenders, he set up specialist committees to look into different areas of scientific interest, he was largely responsible for getting the Society out of its quarters in Gresham College and into a house of their own in Crane Court (off Fleet Street), and he proposed the idea of employing specialists in different areas of scientific research of interest to the Society – an important step away from the notion that all of science could be adequately comprehended by one man, an idea which essentially died with Hooke. Newton was not the sole, pivotal figure who transformed the Royal Society into a professional body; but he was one of the most important of the early presidents, certainly the most important since Brouncker, and he was also the first President whose entire scientific career had been carried out during the lifetime of the Royal.

Less positively, although Newton obtained the appointment of Royal Society 'visitors' to oversee the work of the new Royal Observatory at Greenwich, he also carried on an unedifying row with John Flamsteed, the first Astronomer Royal, who had been commissioned by the King to produce a star catalogue, and was being tediously slow (in the view of everyone except Flamsteed and his wife) in producing the goods. But Newton wouldn't have been Newton if he hadn't been rowing with somebody (the Leibnitz business also continued to drag on). In 1710, he seems to have carried out a last piece of spiteful revenge on Hooke. That year, the Royal was at last able to move out of its cramped quarters in Gresham College to the house at Crane Court. Shortly before the move, a visitor to Gresham commented that

> Finally we were shown the room where the Society usually meets. It is very small and wretched and the best things there are the portraits of its members, of which the most noteworthy are those of Boyle and Hoock.[11]

During the move, overseen with his usual ruthless efficiency by Newton, only one of the portraits got lost – that of Hooke. It has never been seen again, and no portrait of him survives.[12]

11. Conrad von Uffenbach, *London in 1710*.

12. The historian Lisa Jardine has convinced herself that a portrait usually identified as being of the naturalist John Ray, now in the Natural History Museum in London, is actually the missing portrait of Hooke; we are not convinced by the claim.

But there was another side to Newton, who, as we have mentioned, was always generous to his poor relations. One result of this gives a fascinating insight into Newton's world in London in the early years of the eighteenth century. One of those poor relations, his niece Catherine, came to London to be his housekeeper (in the grand sense of running the house; not doing the cooking and dusting, which were jobs for servants).

Catherine (known as Kitty) was the daughter of Newton's half-sister Hannah Smith, who had married a Robert Barton. She was born in 1679, and her father died in 1693, leaving her family destitute. Newton settled all their financial troubles, and in the mid-1690s, when she was about 17, Catherine Barton came to look after his house (then in Jermyn Street, later in St Martin's Street, just south of present-day Leicester Square).

Catherine was a strikingly good-looking, charming young woman who was a great asset to Newton, and she naturally came to know all his circle of friends, including Lord Halifax (who was some twenty years younger than Newton). She was also a close friend of Jonathan Swift, must have known Newton's scientific associates such as Edmond Halley, and was referred to as 'pretty witty Kitty'. Rémond de Montmort, a Frenchman who visited London to observe an eclipse of the Sun in 1715, wrote later from Paris to thank Newton for presents 'given by Mr Newton and chosen by Mrs Barton whose wit and taste are equal to her beauty'. And Kitty was also immortalized in a set of verses composed by an anonymous member of the Kit-Kat Club, a political and literary club that was active in London from about 1700 to 1720. The club took its name from that of a tavern where it first met, which was run by a Christopher (Kit) Cat (or Kat), and was famous for its pies known as 'Kit-Kats'. The membership of around four dozen included Sir Robert Walpole, William Congreve, John Dryden, and Sir Godfrey Kneller, who painted a series of portraits of his fellow members. One of the other members wrote:

> At Barton's feet the God of Love
> His arrows and his Quiver lays
> Forgets he has a Throne above
> And with this lovely Creature stays

Not Venus' Beauties are more bright
But each appear so like the other
That Cupid has mistook the Right
And takes the Nymph to be his mother[13]

Clearly, Kitty Barton was something special – and Lord Halifax was completely smitten. In 1706, when he drew up his will, Halifax added a codicil in which he left £3,000 and all his jewels to Catherine Barton, 'as a small Token of the great Love and Affection I have long had for her'. In October 1706, Halifax purchased an annuity for Catherine, providing her with £200 a year for the rest of her life. And in February 1713, he replaced the codicil in her favour by one giving her £5,000 and the Lodge of Bushey Park to use in her lifetime 'as a Token of the sincere Love, Affection, and Esteem I have long had for her Person, and as a small Recompence for the Pleasure and Happiness I have had in her Conversation'.

To put the bequest in perspective, Halifax's total estate was worth £150,000 when he died in 1715, and Newton received a bequest of £100. Flamsteed, by now Newton's bitter enemy, took great delight, when the will was made public, in writing to a friend that Newton's niece had been well rewarded 'for her *excellent conversation*' (his emphasis). Catherine received the financial part of the bequest, but not the use of the Lodge, which actually belonged to the Crown. Two years later, she married John Conduitt; he was 29, and she was 38. After about 1725, Conduitt largely took over Newton's work at the Mint, and on Newton's death he managed to secure the Mastership formally for himself. It was through the Conduitts, and their daughter (also Catherine), that Newton's papers were preserved largely intact, and eventually found a home in the library at the University of Cambridge.

Having plucked his favourite niece from the prospect of poverty in the provinces, and having been a farmer's boy himself, Newton would surely have been delighted to see his grand-niece Catherine, helped in no small measure by the leg up the social ladder that he had given her mother, marry the Honourable John Wallop, Viscount Lymington, in

13. The verses can be found in *Miscellany Poems* (5th edition, 1727), attributed to John Dryden; but they are almost certainly not from his pen.

1740 – and to see their son, his great-grandnephew, become the second Earl of Portsmouth.

In his declining years, Newton's circle of friends included Prince George (later George II) and his wife Caroline, whom he visited regularly (Queen Anne had been succeeded by the absentee King George in 1714, who was succeeded by George II in 1727). He was rich (when he died, his estate was worth nearly £32,000, excluding the value of his land), famous, and successful, and he was known as a generous benefactor of his extended family and of charity. His books had been reprinted in various editions, and his place in posterity was assured. But, like all of us, he was, in the end, a mere mortal. In 1725, because of illness he moved out of London to the village of Kensington, where the air proved better for him and the Conduitts looked after him with loving care.

But on Christmas Day that year he was 83 years old, and the improvement could not last long. He died on 20 March 1727, in severe pain from a stone in the bladder. At the very last, knowing that death was imminent, he refused the sacrament of the Church, the only public acknowledgement he ever made of his Arianism (at least, semi-public; the only witnesses were John and Catherine Conduitt, and the priest).

Sir Isaac Newton was buried in Westminster Abbey, on 28 March 1727. The best description of the occasion was provided by Voltaire, who saw it and said that it was like 'the funeral of a King who had done well by his subjects'.

The analogy with the death of a king is particularly apt, because of the way Newton's contribution to science was seen by later generations as the defining watershed in the development of this new way of thinking about the world. In fact, as we hope we have convinced you, Newton was by no means the isolated genius of legend (and Pope's couplet) who changed the way people thought about the world single-handedly. His work was the culmination of generations of struggle, and it was far from being the end of the struggle.

To extend Voltaire's analogy, it takes only a little imagination to make a comparison with the way simplistic views of English history see the arrival of Henry VII on the scene, two centuries before Newton, as a defining moment in the development of English politics. Before

Henry Tudor, the simple story runs, England was a nation of warring noblemen, with kings often achieving the throne by murder or treachery; with the advent of the Tudors, the rule of law spread, and even the death of the last Tudor, Elizabeth I, did not result in an outbreak of war, but a peaceful handover to the Stuarts. Just as Tudor apologists such as William Shakespeare portray Henry VII (actually a nasty piece of work) as triumphing over a crook-backed rival, Richard III (who was by no means the nasty piece of work portrayed by Shakespeare), and establishing the rule of law in politics, so Newtonian sycophants portrayed their hero (who was actually a nasty piece of work) as triumphing over a crook-backed rival (Robert Hooke, who was by no means the nasty piece of work they portrayed) and establishing the rule of law in science.

As always, history is written by the victors – or, in both these cases, the survivors. The traditional image of Newton should not be taken as implying a major shift of understanding about the nature of the Universe taking place through the work of one man, or even of one generation; but it is convenient to see the world of science 'before Newton' and 'after Newton' as a shorthand for what happened in science in England in the seventeenth century. Just as the image of what they thought of as Francis Bacon's contribution to science inspired the founders of the Royal Society, so the image of what they thought was Newton's contribution to science (with rather more justification than in the case of Bacon) inspired later generations of scientists in their investigation of the natural world.

Which is why the greatest scientist of the first half of the twentieth century, the man who has become the very symbol of science in modern times, made Newton's pre-eminence clear, in tones which brook no argument. In a foreword to a twentieth-century edition of Newton's *Opticks*, Albert Einstein wrote:

> Nature was to him an open book, whose letters he could read without effort. The conceptions which he used to reduce the material of experience to order seemed to flow spontancously from experience itself, from the beautiful experiments which he ranged in order like playthings and describes with an affectionate wealth of detail. In one person, he combined the experimenter, the theorist, the mechanic and, not least, the artist in exposition. He stands

before us strong, certain, and alone; his joy in creation and his minute precision are evident in every word and every figure.

The particular word we would object to in that eulogy is 'alone'. What can conveniently be called the Newtonian revolution began before him, and carried on after him. Indeed, the defining triumph of the new philosophy came late in the eighteenth century, after Newton and everyone else involved in the story we have told so far was dead. This icing on the cake of the scientific revolution is forever linked with the name of Edmond Halley, who will always be in the shadow of Newton, but who did far more than what 'everybody knows' about him.

*Edmond Halley*

# 8

# The Icing on the Cake

The Big Idea that had been developed in the seventeenth century, from the work of William Gilbert through Francis Bacon and the Fellowship of the Royal Society to be expressed so clearly by Isaac Newton, is now known as the scientific method. It says that the way to proceed when investigating how the world works is first to carry out experiments and/or make observations of the natural world. Then, develop hypotheses to explain these observations, and (crucially) use the hypotheses to make predictions about the outcome of future experiments and/or observations. After comparing the results of those new observations with the predictions of the hypotheses, discard those hypotheses which make false predictions, and retain (at least, for the time being) any hypothesis that makes accurate predictions, elevating it to the status of a theory.

Note that a theory can never be proved right. The best that can ever be said is that it has passed all the tests applied so far. It is always possible that a new kind of experiment might produce a result in conflict with the predictions of a particular theory, so that the theory has to be improved upon, or discarded in favour of a better theory. But when that happens to a long-established theory, it usually turns out that the new theory includes the old one within itself, so that the old theory can still be applied in areas where it does work. This is what happened when Albert Einstein's general theory of relativity subsumed Isaac Newton's theory of gravity. The general theory gives the same 'answers' as Newton's theory for problems like calculating the fall of an apple or the flight of a cannonball on Earth, but it is more accurate when predicting what happens to objects moving in very strong gravitational fields. So nobody bothers to use the general theory for everyday calculations concerning events on Earth (or even for calculating the parameters of the flight of a spacecraft to Mars), where Newton's theory is entirely adequate.

We used gravity as an example here because Newton's theory of gravity (which is essentially a combination of the inverse-square law and the laws of motion) is the archetypal scientific theory. Strictly speaking, however, at the time Newton died it was still merely a hypothesis. It explained everything that was known about the orbits of the planets and moons of the Solar System, the fall of an apple from a tree, or the flight of a cannonball on Earth. But it had yet to make a new prediction that could be tested by observations to see if the hypothesis deserved elevating to the status of a theory. That is where Edmond Halley comes back into our story.

The first thing to get straight about Edmond Halley is his name. In the late seventeenth and early eighteenth centuries, spelling was to some extent a matter of personal taste, and different variations on both his first name and his surname survive in different documents; but the version he used himself on official documents (including his will) was definitely Edmond Halley, and we shall stick with that. But the variations in the written versions of his surname are a useful guide to how that name was pronounced. Many people (including Samuel Pepys) just wrote the name down the way it sounded, without worrying about consistency in spelling at all, and from these phonetic variations on the theme it is clear that it was pronounced 'Hawley', so that the 'Hall' part of the surname rhymes with 'ball'.

To see how Halley put the icing on the cake of Newton's theory, completing the scientific revolution in which the Royal Society played so prominent a part, we need to look back briefly to the 1540s, when Nicolaus Copernicus (1473–1543) published the idea that the Sun, not the Earth, was at the centre of the Universe. Astronomy promptly took a step backwards with one aspect of the work of Tycho Brahe (1546–1601), who suggested that while all the other planets move around the Sun, the Sun moves around the Earth, putting the Earth back at the centre of the Universe. But although this seems now to have been a retrograde step, Tycho took astronomy forward by making a series of superb observations of the movements of the planets across the sky, and these data were inherited by his assistant and successor as Imperial Mathematician to Rudolph II in Prague, Johannes Kepler (1571–1630).

Kepler had made his name at the end of the sixteenth century, when (using the Sun-centred Copernican system) he came up with what seemed a really neat idea, that the invisible spheres to which the six known planets were thought to be attached could be nested inside one another, just at the right relative distances, if they were fitted on either side of a nested set of the five 'perfect solids' (the tetrahedron, the cube, the octahedron, the dodecahedron, and the icosahedron), like nested Russian dolls. It made a kind of sense – if that was the way God liked things, and if the orbits of the planets around the Sun were perfectly circular. There was also a prejudice (going back to the Greeks) that the orbits had to be circular, because circles are perfect, and only perfection is good enough for the heavens.

But when Kepler got to grips with Tycho's data (especially the data for Mars), he found that planetary orbits are not circular; they are elliptical. He found the laws, which we have already alluded to, that describe the motion of a planet around the Sun (for example, that a planet moves faster when it is in the part of its orbit closer to the Sun, in just such a way that an imaginary line stretched from the Sun to the planet sweeps out equal areas in equal times), and began to change the view of the Universe from something set up at the whim of God to something obeying natural laws.

Between them, Kepler and Galileo set the scene for the entry of Isaac Newton. The most important thing about Newton's law of gravity, we reiterate, is not that it is an inverse-square law, interesting though that is, but that it is universal.

This is where Halley, a slightly younger contemporary of Newton, comes in. As we have seen, Halley was instrumental in getting Newton to publish his ideas. Halley also used Newton's law of gravity to predict the return of the comet which now bears Halley's name. The return of Halley's comet in 1759, confirming Halley's prediction and Newton's laws, was the moment when science came of age. From then on, nobody could doubt that the world (what we call the Universe) really is governed by universal laws which are describable in terms of mathematics.

But there was much more to Halley than the man who funded the publication of Newton's *Principia* and predicted the return of a certain comet. Even by the standards of his day, he was an extraordinary polymath – by turns, scientist and diplomat (almost certainly a spy for the British government), an inventor, friend of royalty, sometime naval captain, and leader of the first scientific expedition made by the Royal Navy, a man who left Oxford University without bothering to take his degree, and returned there in triumph as Savilian Professor of Geometry some thirty years later. The thing Halley is most famous for today, his comet, is widely perceived as being merely a footnote to Newton's work. That is far from the truth, for reasons which should already be clear. But in any case, Halley doesn't deserve to be anybody's footnote. It is only compared with Newton or Hooke that Halley seems like anything other than a scientist of the first rank, one of the key people involved in the revolution that changed forever the way we think about the Universe.

Halley was born, according to his own account, on 29 October 1656 (on the old-style Julian calendar still in use in England in those days; it corresponds to 8 November on the modern Gregorian calendar). We only have his word for it, because no record of his birth has ever been found – the relevant parish records may well have been destroyed in the Great Fire of London in 1666, when Halley was a month short of his 10th birthday. His contemporaries included the musician Henry Purcell, born in 1659, and Nell Gwynne, born in the early 1650s, who became a mistress of Charles II.

Some historians wonder whether Halley got the exact date of his birth right, because one surviving document shows that his father, a prosperous businessman and landlord also called Edmond Halley, only married his mother, Anne Robinson, seven weeks before the birth. We shouldn't necessarily jump to the obvious conclusion, though, because this was during the period when England was ruled by Parliament, and many couples at that time were married first by the civil authorities, and only later (if at all) went through a church ceremony. It may well be that the proximity of the birth of their first child encouraged Edmond and Anne to back up their civil ceremony with religious vows.

Young Edmond had a sister, Katherine, who was born in 1658 and died as a baby, and a younger brother, Humphrey, whose exact birth date is not known, but who died in 1684. As the sketchiness of these details shows, very little is known about his early life, and very little, indeed, about his private life in later years. Fortunately, a great deal is known about his public activities, especially where they are mentioned in the writings of his contemporaries (who included John Evelyn, as well as Samuel Pepys).

What we do know is that the family were well off when Edmond was young. One sign of this is that Edmond Halley senior owned a country house 3 miles from the bustle of central London, in Haggerston, now part of the borough of Hackney, but then a peaceful village. It was there that young Edmond was born. His father also owned a town house in Winchester Street, which has since disappeared under railway lines, and as many as a dozen other houses in the same street, which were rented out. And he owned other property. Edmond senior was an important member of the community, and served as a Yeoman Warder of the Tower of London, an essentially honorary post but one which indicates his place in society. The first sign of a decline in the family fortunes came with the Great Fire in 1666, which destroyed some of Halley senior's property, and reduced his income from rents. But his businesses (soap-boiling and salting) flourished in the late 1660s, and there were ample funds available for Edmond to be given the best available education, first at St Paul's School in London, and then at Oxford.

Halley obviously did well at school, where he was appointed School Captain in 1671 and where he developed an existing boyhood interest in astronomy. When he went up to Queen's College, in Oxford, on 25 June 1673, he had a good knowledge of Latin, Greek, and Hebrew, was a more than competent mathematician, knew the basics of navigation, and had developed some skill as an observational astronomer. His affluent background was a great help in pursuing his astronomical interests, and he arrived in Oxford with a set of instruments, including a telescope 24 feet long and a sextant 2 feet in diameter, that would have been the envy of many a contemporary professional astronomer.

There was one cloud on the horizon, however. Halley's mother died in 1672. We don't know exactly where, when, or why; but she was buried on 24 October that year. In due course, as we shall see, this was to have a profound impact on Edmond when his father made an unfortunate second marriage.

While Halley was growing up, science (and in particular astronomy) was taking off in England, with the founding of the Royal Society in the heady years of the Restoration. The need for better methods of navigation for the growing navy led the science enthusiast Charles II to appoint John Flamsteed (1646–1719) as his 'astronomical observator' in March 1675. One of his key tasks would be to carry out a new survey of the northern skies, using telescopic methods to improve upon the best existing catalogue, made by Tycho using open sights. The post metamorphosed into that of Astronomer Royal, and later the same year the King ordered the construction of the Royal Greenwich Observatory, under the direction of Wren and Hooke, as a base for Flamsteed and his successors. The Royal Warrant, dated 22 June 1675, that established the observatory begins:

> Whereas, in order to the finding out of the longitude of places for perfecting navigation and astronomy, we have resolved to build a small observatory within our park at Greenwich, upon the highest ground, at or near the place where the castle stood, with lodging rooms for our astronomical observator and assistant, Our Will and Pleasure is that according to such plot and design as shall be given you by our trusty and well-beloved Sir Christopher Wren, Knight, our surveyor-general of the place and scite of the said observatory, you cause the same to be built and finished with all convenient speed.

Just a week after Flamsteed's appointment, Edmond Halley, then an 18-year-old Oxford undergraduate, wrote to the King's new 'observator' describing some results of his astronomical observations, which seemed to suggest that published tables of astronomical data were wrong, and politely asked if Flamsteed could confirm that Halley's numbers were correct. By the summer, Halley was making observations with Flamsteed, had met Hooke on his visits to London, and was described by Flamsteed (in the *Philosophical Transactions*) as 'a talented young man of Oxford'. It was in no small measure through Flamsteed's support and encouragement that Halley wrote his first scientific paper (about planetary orbits) later in 1675, while still an undergraduate. It was published the following year. Two other papers quickly followed.

Halley was, by all accounts, an outstandingly able student, and would surely have passed his degree with flying colours. But he became impatient to build on the success of his early work and to make a name for himself in the way that people like Flamsteed were doing. Flamsteed was making a catalogue of the positions of the stars, using the new technology of the telescope, which made his measurements far more accurate than those of older catalogues obtained with more primitive sighting instruments – using 'open sights'. Halley hit on a scheme of doing the same kind of thing for the stars of the southern sky, from the outpost of St Helena, in the Atlantic, at that time the most southerly British outpost. As he wrote to Oldenburg, he had an idea to go

> To St Helena or some other place, where the south pole is considerably elevate, by the next East Indian Fleet, and to carry with me large and accurate instruments, so as to be able to make a most accurate sphere of the fixed stars, and complete our globe throughout.

With his father's influence in high places and the support of Brouncker, as well as Flamsteed, the scheme was put to the government and to the King, who 'recommended' to the East India Company that the young man and a friend, James Clerke, be given passage to St Helena, which the company controlled. His father supported the idea to the tune of offering Halley an allowance of £300 a year (which was three times Flamsteed's salary as Astronomer Royal). They sailed in November 1676, the degree, for the moment, forgotten. Halley was just 20.

With severe weather to contend with, the catalogue took more than a year to complete. Halley also observed a transit of the planet Mercury across the Sun's disc, on 28 October 1677, and carried out other scientific studies while he was there. But the trip also gave rise to a bizarre rumour about Halley's private life. Hints of sexual impropriety followed Halley throughout his early adult life, starting with this visit to St Helena. John Aubrey, in his *Brief Lives* (not published until long after Aubrey's death), alludes to a long-married but childless couple who travelled out to St Helena in the same ship as Halley, and remarks that 'before he came [home] from the Island, she was brought to bed of a Child'. The implication was that Halley was the father, and it has been repeated down the centuries. The truth is that when Halley had returned to London in 1678 he had told, as an example of the benefits of sea air, the tale of a man of 55 with a wife of 52 who had sailed on the same ship with him to settle on St Helena, and were expecting a child by the time he left the island.

Halley's return from the South Atlantic in the spring of 1678 was something of a triumph; his *Catalogue of the Southern Stars* was published in November that year, earning him the sobriquet 'Our Southern Tycho' from Flamsteed himself, and his results were described to the Royal by Hooke. There were limitations to the survey – it was a quick job covering only the most important stars, and their positions were determined relative to the old catalogue of Tycho, from the parts of the sky where the two surveys overlapped. It could not be as accurate or complete as the new survey of the northern sky being carried out by Flamsteed. But it was, nevertheless, not only the first survey of the southern stars, but the first star survey to use modern telescopic methods. And the positions of the stars in Halley's catalogue could always be corrected once Flamsteed's observations were complete, because Flamsteed was studying many of the same stars as Tycho. It's an indication of Halley's new status that, as well as making observations from time to time with Flamsteed, when Flamsteed was ill in the summer of 1678, Halley carried out observations from Greenwich on his behalf.

While he was at St Helena, Halley, as we have mentioned, was also lucky enough to observe a rare transit of Mercury, when the planet passes in front of the Sun's disc as seen from Earth. This drew his attention to the fact that observing such transits of Mercury and Venus (the only two planets closer to the Sun than we are) can provide a measure of the distance to the Sun, using geometrical techniques and careful timing of the exact moments when the planet 'enters' and 'leaves' the solar disc. The observations made in 1677 were not good enough for the calculation to be carried out, but it started Halley on a path which would eventually lead to the first accurate measurement of the distance to the Sun.

There was also another small matter that needed to be put right. Halley's work was clearly more than good enough to justify him being awarded a degree, but the rules at Oxford were strict – if a student stayed up at college for a certain number of terms, he would get a degree more or less regardless of his intellectual merits; but Halley had not stayed up for the minimum number of terms required, which meant no degree, no matter how good his intellectual merits. At the suggestion of Sir Joseph Williamson, the Secretary of State (who had been instrumental in obtaining approval for the St Helena expedition), the King wrote to the Vice-Chancellor of Oxford with a 'recommendation' that could not be ignored:

> Whereof [Halley] has (as We are informed) gotten a good Testimony by the Observation he has made during his abode in the Island of St Helena; We have thought fit for his Encouragement hereby to recommend him effectually to you for his Degree of master of Arts; Willing and Requiring you forthwith upon the receipt hereof (all Dispensations necessary being first granted) to admit him the sd. Edmund Hally to the said Degree of Master of Arts without any Condicion of performing any previous or subsequent Exercises for the same any Statute or Statutes of that Our University to the contrary in any Wise notwithstanding.

Halley received the degree on 3 December, just three days *after* being elected a Fellow of the Royal Society. His next task was to visit the German astronomer Johannes Hevelius (1611–87), in Danzig, as a representative of the Royal itself.

This was a mission requiring tact. The problem was that Hevelius, although a superb astronomer in his day, was by now an old man who clung to the old 'open sights' method of making observations of star positions, but claimed incredible accuracy for his results. The younger generation of astronomers, using telescopic methods, were sceptical; Hooke in particular had been highly critical of the claims made by Hevelius. The Royal wanted to set the matter straight – not out of malice, but for scientific reasons. Could the old astronomer's data be trusted or not? Halley went to find out. It isn't entirely clear whether he made the trip to Danzig at the instigation of the Royal, or whether he had planned such a trip anyway, and the Royal took advantage of the opportunity to use him as their emissary. It is, though, worth mentioning that there was a thriving trade between Britain and the Baltic in those days, and that Danzig was one of the principal Hanseatic ports. Much of London's wealth was based on overseas trade, and many of the merchants in the circles in which Edmond Halley senior moved would have been familiar with the port. It was as easy for Halley to get there as to almost any city in Europe.

Halley reached Danzig in May 1679, and carried out a long series of observations with Hevelius and his colleagues. He reported back in letters to Flamsteed that, indeed, everything was fine, and Hevelius really had been achieving the accuracy he claimed. But this seems to have been a tactful assessment for public consumption, while in private he was more critical. On 14 August, Hooke noted in his diary: 'Halley returned this day from Dantzick (Hevelius Rods in pisse).' This can only mean that Halley had been privately dismissive of the results obtained by sighting along 'Rods'. In public, still courteous, Halley never actually endorsed Hevelius' claims, but was less critical than might have been expected.

In a letter to a friend, written in 1686 (when Hevelius was 75), Halley referred to Hevelius as 'an old peevish Gentleman, who would not have believed it possible to do better than he has done', and in a subsequent letter the same year commented: 'I am very unwilling to let my indignation loose upon him, but will unless I see some publick notice elsewhere, let it sleep till after his death if I chance to outlive him, for I would not hasten his departure by exposing him and his observations as I could do and as I truly think he deserves I should.'

This curious reluctance of Halley to condemn Hevelius publicly lent some weight to another scurrilous rumour that grew out of his trip to Danzig, although there is no evidence that the story is true. The story grew, as such stories do, out of the age difference between Hevelius and his wife. Hevelius' first wife had died in 1662, and in 1663, the year he was 52, he had married Elizabetha, the beautiful 16-year-old daughter of a rich merchant. When Halley visited Danzig in 1679, Elizabetha was still only 32, and a known beauty; Hevelius was 68; Halley was a vigorous young man in his 23rd year; and the rest is rumour and innuendo (but plenty of it, at the time and subsequently).

Whatever the reason, the fact that Halley initially reported favourably on Hevelius' open-sights observations, and that it soon became clear that they were not as good as he claimed, contributed to a rift between Halley and Flamsteed which was never made up. It is possible that Flamsteed, a very serious-minded man, took exception to the rumours about Halley and Elizabetha Hevelius, whether they were true or not; and the continuation of the quarrel no doubt owed much to the way in which Flamsteed's former protégé soon began to outdo the Astronomer Royal himself.

But Halley did not hasten to develop his career in science. Still comfortably supported by an allowance from his father, for just over a year he divided his time between Oxford and London, attending meetings of the Royal Society and (we know from the papers of his contemporaries) visiting one of the fashionable coffee houses, Jonathan's (also a favourite haunt of Robert Hooke), in Change Alley. At the end of this period, in the winter of 1680–81, the bright comet we mentioned earlier was seen rounding the Sun, and was a major talking point, not just among astronomers. First seen in November 1680, the comet was observed moving towards the Sun, and being lost in its glare; a little later, it was seen moving away from the Sun. It became the brightest comet that anyone alive at the time had witnessed, clearly visible from the streets of cities like London and Paris in those days before artificial lighting.

But this was before Newton published his work on gravity, and most people thought that what they had seen was two different comets, one moving towards the Sun and another moving away. Comets were still mysterious and inexplicable heavenly phenomena; and although Flamsteed was one of the few people who suggested that this might have been one object moving first towards and then away from the Sun, as we have mentioned he had no inkling that it was in an orbit around the Sun, and thought that its motion had been reversed by magnetic repulsion.

Halley saw the 'first' comet from London at the end of 1680, just before he set off to do the Grand Tour of Europe, accompanied by a friend, Robert Nelson, the son of another rich City businessman. This tour of France and Italy, featuring visits to Paris and Rome, was the done thing for rich young gentlemen (and, indeed, ladies), but in Halley's case it would enable him to meet astronomers and other men of learning on the continent. One of the main talking points among both scientists and other people in the early part of their journey was the 'second' comet; and among the people Halley discussed it with was Giovanni Cassini, the head of the Paris Observatory, who was the first person to suggest that the rings of Saturn are made up of a myriad of tiny particles, orbiting Saturn like miniature moons.[1] They carried out observations of the comet together, and like many contemporary astronomers tried unsuccessfully to work out orbits for comets and find ways to predict their return. In a letter to Hooke dated 15 January 1681, Halley wrote:

The generall talk of the virtuosi here is about the Comet, which now appears, but the cloudy weather has permitted him to be but very seldom observed. Whatever shall be made publick about him here, I shall take care to send you, and I hope when you shall please to write to me you will do me the favour to let me know what has been observed in England.

From Paris the tour took in many of the cities of France and Italy on its way to Rome, where Nelson stayed and found love, marrying the splendidly named Theophile, daughter of Sir Kingsmill Lacy and widow of the first Earl of Berkeley. Another resident of Rome at the time was the exiled Queen Christina of Sweden (1626–89), who also took an interest in the comet, and offered a prize to anyone who could predict its future path; the prize was never claimed. Halley came back to England by an even more circuitous route than he had travelled out to Rome, taking in Holland as well as Paris, where the Palace of Versailles was just being completed. Flexing his muscles, Louis XIV (the 'Sun King') was already beginning the harassment of French Protestants that would lead to an exodus of Huguenots (including Abraham de Moivre) to England. In Paris, Halley had developed another of his interests, making a comparison of the sizes of the populations of Paris and London, studying the records of births and deaths, and calculating that for the population of Paris to remain stable each couple would need to have four children. He was back in London on 24 January 1682, having, in the words of Aubrey, 'contracted an acquaintance and friendship with all the eminest mathematicians in France and Italy'.

At this point, our lack of detailed knowledge of Halley's private life

1. On his return to England, early in 1682, Halley discussed his observations of the comet with Newton, the first known meeting between the two.

becomes frustrating. Although his trip around Europe seems leisurely by today's standards, Halley actually spent only a few weeks more than a year abroad, quite short for a Grand Tour in those days. We know that his father remarried at about this time, although we don't know exactly when, and it is possible that Halley hurried back to attend the wedding. Whatever, on 20 April 1682 (and completely out of the blue as far as any surviving historical records are concerned) Halley, now 25, himself got married, to Mary Tooke, the daughter of Christopher Tooke, a lawyer at the Inner Temple. The wedding took place at St James's Church, in Duke Place, not far from the Winchester Street home of Halley's father.

Marriages in those days, especially those involving members of wealthy and influential families, were seldom made for love (at least, not for love alone), and although the details have been lost it is likely that Halley's marriage was arranged as one that would be of benefit to both families. His cousin, Anne Cawthorne, also married about this time, to a merchant who was related to Edmond Halley senior's new wife; Alan Cook has argued that the three marriages may have been interconnected in some way by business interests, and that this would explain Halley's sudden return from France.

But all this is speculation, albeit informed speculation. So little is known about Halley's married life that we might as well give you all the known details of the marriage here and now. The couple were together for more than fifty years, and seem to have been very happy with one another for all that time. They had three children: Edmond (who was born in 1698, became a surgeon in the navy, married Sybilla Freeman, and died, childless, shortly before his father), and two daughters – Margaret (born in 1685, who never married) and Katherine (born in 1688, who married twice but left no children). There may have been other babies who died in infancy (there usually were, in those days). And that's it.

Edmond and Mary Halley set up home in Islington, then a village just to the north of London, where he established a small but well-equipped private observatory. Halley, still funded by his father, was an active observer, publishing papers on the motion of the Moon, Saturn, and one of its moons, Rhea, and on the tides; he also played an important part in the activities of the Royal Society. He was also interested in the idea that variations in the magnetism of the Earth from place to place could be used as an aid to navigation – indeed, he had made magnetic observations on his voyage to St Helena and back. And he was also keenly interested in another bright comet that appeared in the skies in 1682.

In the outside world, 1682 was the year when William Penn, a Fellow of the Royal Society and a friend of Halley, set out to found the Pennsylvania Colony in North America. Overall, this was the most intense period of cold of what has become known as the Little Ice Age; in the winter of 1683–4 the Thames at London was covered with nearly a foot of ice. We mentioned earlier that a 'Frost Fair' was set up on the frozen river that winter; it was the record of this winter that formed the basis for the severe winter weather described much later in *Lorna Doone*.

At the end of 1683, Halley seemed secure and successful. He had been elected to the Council of the Royal Society that year, he had his own private observatory, no financial worries, and was settled in his marriage. But just as that severe winter was ending, his life changed. On Wednesday, 5 March 1684, his father went out and never returned. Five days later, his stripped body was found by the river at Temple Farm. The resulting inquest came up with a verdict of murder, but no culprit was ever found, and no reason for the murder was officially established. One possibility is that the elder Halley had been involved, in his capacity as a Yeoman Warder, with the machinations during the imprisonment in the Tower and suicide of the Earl of Essex, awaiting trial for high treason. Essex had been involved in a desperate scheme, the Rye House Plot, to remove any prospect of a Catholic succession by assassinating both Charles II (suspected, rightly, of Catholic leanings) and his openly Catholic brother James, the heir to the throne. There is little doubt that Essex committed suicide, using a razor lent to him by a guard to trim his fingernails. But his death before trial was convenient for the authorities, and the guard seems to have been remarkably careless. A story current at the time, and entirely plausible, is that Edmond Halley senior knew too much and didn't have enough sense to keep quiet about what he knew, so had to be silenced.

Whatever the cause of Halley's father's death, it had a dramatic effect on the younger Edmond's life. His father had died intestate, and this led to prolonged legal wrangling between the son and the widow, Joane Halley. It is sometimes suggested that Halley ended up in a bad way financially, but this seems to be an exaggeration. Certainly the older Edmond Halley's fortune had been diminished by this time, partly because of the loss of income following the Great Fire, and partly because of the extravagance of his second wife, but what was left (not least, all the houses in Winchester Street) was certainly worth fighting for. The younger Edmond Halley did not become a pauper overnight (in any case, almost certainly, his wife had brought a reasonable dowry with her), but it is true that he could no longer rely on the secure and generous financial support he had had throughout his career. But his brother Humphrey had already died somewhere overseas, so there were only two possible heirs to share what was still a considerable inheritance. The situation was complicated by Joane's remarriage, to one Robert Cleeter, in 1685; but Halley ended up with all the houses in Winchester Street, except the one occupied by his stepmother, and some of his father's financial investments. This brought him, altogether, an income of some £200 a year (between £100,000 and £200,000 in today's terms).

But the death of his father had an impact far beyond the Halley family. It was because Halley had to attend to family business at Alconbury, near Huntingdon, in connection with settling his father's estate, that Halley visited Isaac Newton, in Cambridge, in 1684.

Newton and Halley had already corresponded about the comet of 1682, and it is likely that they met briefly that year on one of Newton's visits to London. At that time, though, Newton was still keeping himself to himself, and took little part in the activities of the Royal. What brought Halley and Newton together in 1684, as we have seen, was the growing interest among many astronomers in *why* the orbits of the planets around the Sun should obey Kepler's laws.

Halley became interested in the problem. He realized that the numbers in Kepler's laws, relating the time it takes a planet to go round the Sun to the distance of that planet from the Sun, could be explained if there was a force of attraction between the Sun and the planet which could be calculated as one over the square of the distance between them. It was on the evening of 14 January 1684 that the famous conversation on the subject took place between Halley, Hooke, and Wren, and Halley later wrote to Newton (in 1686) that:

> This I know to be true, that in January 83/4, I having from the sesquialtera proportion of Kepler, concluded that the centripetall force decreased in the proportion of the squares of the distances reciprocally, came one Wednesday to town, where I met with Sr Christ. Wren and Mr Hook, and falling into discourse about it, Mr Hook affirmed that upon that principle all the Laws of the celestiall motions were to be demonstrated.

We have seen how this led to the meeting with Newton described by de Moivre, and how Newton rummaged in his papers, but claimed not to be able to lay hands on the proof. He may just have been playing for time, still cautious about going public with his ideas and risking another public argument with the likes of Hooke. But he promised to work out the proof again, and send it on to Halley in London.

In two months' time, Halley held in his hands the nine-page paper from Newton which elegantly explained the mathematical basis for all of Kepler's laws, using the inverse-square formulation of the law of gravity. It was clear that this was just the tip of an iceberg of unpublished work, and Halley went back to Cambridge to find out just how much Newton had still hidden up his sleeve. It was on this trip that he persuaded Newton to write what became his famous book, the *Principia*. In it, Newton would explain how gravity not just holds the planets in their orbits around the Sun, but holds the Moon in orbit around the Earth, and explains the fall of an apple from a tree. He would put forward his three laws of motion, and lay the basis of a system of mechanics that would be unchallenged until Albert Einstein came along, and which, as we have seen, is still the basis of everyday needs, used, for example, to calculate the construction of a bridge or the flight of a space probe to its intended destination.

Halley reported back to the Royal on 19 December 1684, with the news that Newton had promised to publish his results, and would send the mathematical proofs to the Society in due course. It took Newton well over a year to complete his epic, constantly encouraged by Halley, who provided comments on the various drafts as the work progressed, and eventually took on the responsibility for reading the proofs.

It was during this period that Halley's own situation changed, and on 7 February 1686 he became Clerk to the society. He had already moved house, to Golden Lion Court,[2] in Aldersgate Street, in 1685; so he was conveniently located for the Society's meetings. Nevertheless, at first sight this seems a strange career move, which has been interpreted by some people as implying that he was in need of the money provided by the modest salary (notionally £50 a year, but seldom paid in full by the impoverished Society) that went with the position; but, as we have seen, this seems unlikely. One of the stipulations of the post was that as a paid employee of the Royal the holder could not be a Fellow, so Halley had to resign his Fellowship (which makes the move look even more strange). But other stipulations, in particular the requirements that the Clerk should be a single man and live in Gresham College, were waived or ignored. Clearly, the Society was as keen for Halley to be Clerk as he was to hold the post. The truth seems to be that Halley was eager to get more involved in the affairs of the Society, and they were eager to have such an able young man (he was still only 29) on board. Far from being reduced to taking the post by poverty, it appears that Halley was able to take on the role precisely because he had independent means and didn't need the largely notional income, and that this was yet another reason why he suited the Society. Since he also took on the financial responsibility for the publication of the *Philosophical Transactions* during his tenure as Clerk, he can hardly have been hard up.

Whatever his reasons for taking up the post (and the expectation of an income seems to have been the least of them), this was to be good news for Newton. Halley was a diligent and efficient Clerk, who did much to ensure the smooth running of the Society's affairs, while continuing with his own research. The first part of the *Principia* was delivered to the Royal in April 1686, and although the details are vague it seems clear that Halley was instrumental in ensuring that on 19 May the Society ordered that it should be printed at the Royal's expense (incidentally, the President of the Royal at the time was Samuel Pepys).

Unfortunately, the Society was not able to live up to its promise.

2. This house may also have been part of his inheritance.

A little earlier, they had used the meagre fund at their disposal to pay for the publication of Francis Willughby's *History of Fishes*. It proved virtually unsaleable – there were still copies on the inventory of the Royal in 1743 – and with no money to hand, instead of paying Halley his £50 salary that year the Society gave him fifty copies of the book! There was certainly no money in hand for the publication of Newton's book, and Halley had very little choice but to make good the promise by funding the publication himself. There is, though, a happy ending; unlike the fish book, the *Principia* was a modest commercial success, as well as being a runaway intellectual triumph, sold reasonably well, and Halley made a tiny profit out of it. He certainly earned it. And Newton certainly appreciated the effort he had put in, writing in the Preface to the great work that

In the publication of this Work, the most acute and universally learned Mr Edmund Halley not only assisted me with his pains in correcting the press and taking care of the Schemes, but it was to his solicitations that its becoming publick is owing. For when he had obtained of me my demonstration of the celestial orbits, he continually pressed me to communicate the same to the Royal Society; who afterwards by their kind encouragement and entreaties, engaged me to think of publishing them.

Halley's efforts on Newton's behalf did not end with the publication of the *Principia*. The book appeared in July 1687. A copy of the first edition was presented by Halley to the new King, James II, who had succeeded Charles II in 1685; he also sent specially inscribed presentation copies to Cassini in Paris and to Newton's special friend Fatio de Duillier, among others. He also reviewed the book, in the *Philosophical Transactions*.

This was the beginning of a turbulent time in English politics, as the Catholic James II was shortly (in 1689) replaced by the Protestant William and Mary;[3] we have seen how all this affected Newton, but none of it deeply concerned Halley. This was when he wrote:

3. The turbulence included wars with the Dutch in the 1680s, followed by the Nine Years War (caused by the expansionist policies of Louis XIV of France) in which, after the succession of William and Mary, the English and Dutch were part of a coalition against the French. The war began in 1688, but the English joined in a year later.

For my part, I am for the King in possession. If I am protected, I am content. I am sure we pay dear enough for our Protection, & why should we not have the Benefit of it?

If the rumours about the death of his father were true, Halley had good reason to take such a stance. Such diplomacy would soon be put to good use.

In the late 1680s and early 1690s, while still working diligently as Clerk to the Royal Society, Halley published a series of scientific papers and discussed other ideas at meetings of the Royal. These included mathematical papers (he was always good at maths), various astronomical studies, and an investigation of the possible causes of the biblical flood. This last piece of work was to have an almost immediate adverse effect on his career. In those days, the Church accepted the biblical chronology which, by counting back the generations listed in the Bible from Jesus to Adam, came up with a date of 4004 BC for the creation of the Earth (if you want to be precise, John Lightfoot, Vice-Chancellor of the University of Cambridge, had declared, following the publication of Archbishop Ussher's *Sacred Chronology* in 1620, that Adam had been created at 9 a.m. on Sunday, 23 October 4004 BC). Halley realized that the kind of dramatic changes in the features of the Earth associated with an event on the scale of the Flood implied a much longer history – he accepted that there was some basis in the story in the Bible, but in trying to put this on a scientific footing he ran into conflict with traditional churchmen by suggesting a very much longer age of the Earth than they were willing to countenance. When he tried to estimate the age of the Earth by analysing the saltiness of the sea, he came up with a similarly long timescale of history.

Around this time, in the mid-1680s, Halley's relationship with Flamsteed deteriorated to the point of hostility (at least, on Flamsteed's part), although the reasons are not entirely clear. But this was no more than a background irritation as Halley, now in his thirties, restlessly developed his scientific interests in different directions as the 1680s gave way to the 1690s.

Ever since his voyage to St Helena, Halley had maintained an interest in the sea, and in the workings of the ocean and atmosphere.

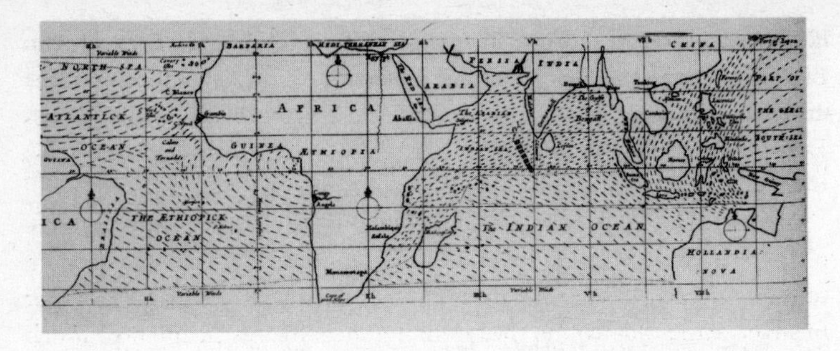


*Halley's chart of the trade winds*

He carried out a survey of the Thames estuary, which was of key strategic importance in any conflict with the navies of continental powers. He was intrigued by the idea of using magnetic variations to measure the position of a ship at sea (which led him to develop one of the first theories of the Earth's magnetism), and he was interested in the way the wind blows. He found the law which relates the change in pressure of the atmosphere to height above sea level, and in 1686 he published a paper on the trade winds and monsoons, which included the first meteorological chart, a map showing the average pattern of the winds that blow around the world. But Halley was also very much a practical man, and at the beginning of the 1690s he was off to Pagham, in Sussex, where he carried out experiments in deep-sea diving, using a frigate provided by the Admiralty to try to salvage gold and other valuables from the wreck of the *Guynie*. There is no record of whether the salvage attempt succeeded, but in the course of this work Halley invented a practical diving bell, which enabled men to work on the sea bed in a depth of 10 fathoms for up to two hours at a time. And, being a good scientist, while he was at it he carried out experiments on the behaviour of sound and light under water.

Alongside all this, as Clerk of the Royal Society for fourteen years (although, as we shall see, he was away for the last four of those years, and others deputized for him) Halley helped to steer the Society through the difficulties which followed the twenty-fifth anniversary of its foundation. It was Halley who established the *Philosophical*

*Transactions* on a secure and (as it proved) permanent basis, and who brought before the Society reports of many different scientific ideas and experiments, as well as communicating with scientists across Europe on their behalf. But, important though all this was, it is his own scientific work that we shall concentrate on here.

In 1691, Halley published a paper giving the first scientific estimate of the size of atoms. He knew how silver could be coated with gold to make silver-gilt wire, by drawing the silver wire out from an ingot of silver with gold around the circumference. By measuring how much gold was used in the process, he calculated that the skin of gold around the final wire was 1/134,500th of an inch thick, and assumed that the layer must be at least one atom thick. From this, he calculated that a cube of gold with sides each one-hundredth of an inch long would contain at least 2,433 million atoms. Of course, Halley realized that the layer of gold on the wire must be more than one atom thick, so that his tiny cube must really contain even more atoms, of correspondingly smaller size. But this was a genuinely scientific attempt to measure the sizes of atoms, at a time when few people really took the idea of atoms seriously.

Halley was also interested in measuring the Universe at large. The same year, he published another paper explaining how observations of a transit of Venus could be used to measure the distance to the Sun. Because of the geometry of the orbits of the Earth and Venus, these transits are rare events, occurring in pairs (with the two transits in each pair about eight years apart) at intervals of more than a hundred years. Halley showed how observers at widely separated places on Earth would see the transit slightly differently, because of parallax (to see parallax at work, hold up one finger at arm's length and watch how it seems to move when you close each of your eyes alternately). By comparing their observations, the two teams could use the geometry of the Earth–Venus–Sun alignment to work out the distance to the Sun.

Halley would return to this theme in 1716, when he predicted that the next pair of transits of Venus would occur in 1761 and 1769, and wrote:

> I strongly urge diligent searchers of the heavens (for whom, when I have ended my days, these sights are being kept in store) to bear in mind this

injunction of mine and to apply themselves actively and with all their might to making the necessary observations.

But a lot of water would flow under the bridge before he published that paper.

Even allowing for the role of amateurs and gentlemen of independent means in science in the seventeenth century, it is amazing that this outpouring of work came from a man who was officially only the Clerk to the Royal Society. Halley himself eventually chafed at this, and became eager to establish himself in a more important academic position. The opportunity seemed to present itself in June 1691, when he was carrying out his diving experiments at Pagham and heard that Edward Bernard, the Savilian Professor of Astronomy at Oxford University had resigned his post. Halley thought that he had a good chance of filling the vacancy, and put his name forward. But there was a snag. The appointment was subject to the approval of the Church authorities, and Halley's views on the age of the Earth had caused him to be regarded as a heretic. As he wrote from Pagham, to a friend, Abraham Hill, he was concerned that he might not get the post, 'there being a caveat entered against me, till I can show that I am not guilty of asserting the eternity of the world'.

He was right. Partly for his heretical views, partly because of the opposition of Flamsteed, and in spite of the support of the Royal Society for his application, he didn't get the job, which went to David Gregory (1661–1708), a nephew of the James Gregory we met earlier, and one of Newton's protégés. Gregory had previously been a professor in Edinburgh. At the time, an established academic with a sound reputation in mathematics must have looked a better candidate than someone who had never held a university post and seemed to be more interested in deep-sea diving than astronomy; the religious doubts, though real, were probably a secondary consideration.

It looked as if the doors of academe were closed against Halley. But this didn't slow down his scientific work, which continued in all its diversity. It was in March 1693, for example, that he presented his calculations of human mortality, the first scientific basis for life insurance, to the Royal Society. His analysis of the records of births and deaths showed how it was possible to calculate the life expectancy of a person according to their present age, and therefore how to determine the rate at which annuities should be paid. And in the mid-1690s Halley corresponded with Newton about the nature of the orbits of comets – a theme we shall return to shortly.

But by that time Halley was already involved in another project that would preoccupy him for most of the rest of the decade, and into the early years of the eighteenth century. Also in 1693, together with a friend, Benjamin Middleton, Halley came up with a scheme for a scientific voyage of discovery. Middleton, who had been elected as a Fellow of the Royal in 1687, was the son of Colonel Thomas Middleton, a Commissioner of the Navy who had been a friend of Samuel Pepys and had been in charge of the dockyards at Chatham and Portsmouth during his career (he had died in 1672). The younger Middleton must have had some knowledge of nautical matters, but his role in the original proposal was largely that of the financial backer for the project, offering to pay for 'Victualls and Wages' if the Admiralty would supply a ship.

Middleton made a formal proposal to the Admiralty requesting that a ship be provided for a voyage of scientific investigation, to circumnavigate the globe and, among other things, to test methods of determining longitude at sea – the key problem in navigation at the time, something of huge importance to the navy, amply justifying the loan of a ship. It is clear that Middleton was at first the leading light in the project, since Robert Hooke's diary records (11 January 1693) that Halley had spoken 'of going in Middleton's ship to discover' (that is, to explore). On 12 July 1693 the Commissioners of the Admiralty wrote to the Navy Board that:

> Her Ma$^{ty}$ is graciously pleased to incourage the said undertakeing. And in pursuance of her Ma$^{ts}$ pleasure Signified therein to this Board, We do hereby desire and direct you forthwith to cause a Vessel of about Eighty Tuns Burthen to be set up and built in their Ma$^{ts}$ Yard at Deptford as soon as may be, and that Mr Middleton be consulted with about the conveniences to be made in her for Men and Provisions, and that when she is built She be fitted out to Sea, and furnished with Boatswains and Carpenters stores for the intended Voyage, & delivered by Inventory to the said Mr Middleton to be returned by him when the Service proposed shall be over.

Although Middleton had the money, Halley was clearly behind the scientific purpose of the expedition. He must also have had even more experience at sea than we know about. His survey of the Thames approaches was not the work of a nautical novice, and as early as May 1688 he had reported to the Royal Society how the sailors on board a guardship anchored 4 miles offshore remained healthy, although the forts at nearby Sheerness were notoriously unhealthy. It seems likely that he had been at sea in 1687 or before, but in what capacity we do not know.

As the letter quoted above indicates, the new proposal, combined with Halley's experience, met with an enthusiastic response and a small ship was specially built, at Deptford, for the proposed expedition. It was of a kind known as a pink, developed in the Netherlands, with three masts, bulging sides, and a flat bottom, making them useful for carrying stores and navigating in shallow waters. The vessel was launched on 1 April 1694, and named the *Paramore*; she was 52 feet long, 18 feet wide at the beam, had a draught of 9 feet 7 inches, and displaced 89 tons. Try pacing those dimensions out in the park and see if you would fancy voyaging to the South Atlantic in her.

After all that, there was a delay of nearly two years before anything more is heard of the project. We have no record of any reason for the delay, but apart from the lengthy process of fitting out the ship it seems at least possible that the continuing war with France meant that it was felt too dangerous, or inappropriate, to send such a small vessel out alone on such a mission.[4] Halley went about his work as Clerk to the Royal Society, and continued to carry out his own scientific studies. By 1696, though, the expedition is clearly Halley's affair. So much so that on 4 June that year he received a Royal Commission from William III (Mary had died on 28 December 1694) appointing him as Master and Commander of the *Paramore*. In other words, Halley, a landsman, had been appointed as Captain of a King's ship, in the Royal Navy.

This was unusual, although it wasn't quite such a dramatic step as

4. In the summer of 1693, a fleet of several hundred English and Dutch merchantmen was intercepted by French warships off Lagos, in the south of Portugal, and suffered heavy losses. It is probable that Halley was among those who suffered financially as a result, since Hooke's diary records 'Hallys trade taken by French' on 24 July that year, the date news of the disaster reached London.

it is sometimes painted. First, Halley was an experienced seaman, even if he had not made a career in the Royal Navy. Secondly, at the end of the seventeenth century the role of commanders of King's ships was still in transition (it was Pepys, of course, who initiated the reforms which led to the establishment of the Royal Navy in essentially the form it retained for the next three centuries). Originally, there was a clear distinction between the officers and gentlemen who were in command, and the sailors who made sure that the commands were carried out. For example, Henry FitzRoy, the Duke of Grafton and an illegitimate son of Charles II, served his father both in the army on land and as an admiral in his ship the *Grafton*. Such commanders were responsible for strategy and tactics, and where necessary diplomacy – an especially important requirement for the commander of a ship in distant waters, where he was the representative of the King and government of England. In that capacity, the rank of Captain in the navy was regarded as equivalent to Colonel in the army. The 'tarpaulins', officers who knew about handling ships under sail, were responsible for the practicalities of getting the ship to do what the Captain (or Admiral) wanted.

You might think that Halley could have sailed as a passenger on the *Paramore*, in the way that Charles Darwin would later sail as a passenger on the *Beagle*. But it made more sense for him to have the rank of Captain. Unlike the *Beagle*, the sole purpose of the *Paramore*'s voyage was scientific exploration. At that time, in 1696, the expedition was seen as being the responsibility of the Royal Society, a private voyage using a ship borrowed from the navy (even though that ship had been specially built for the job). So it was clear that Halley should be in charge. On the other hand, to ensure discipline, it was intended that the crew should be provided by the navy, with the normal chain of command. And that required a commander. What made the appointment unusual, and possibly unique in the history of the Royal Navy, is that Halley wasn't the sort of aristocratic gentleman usually appointed to command – he was certainly no Duke of Grafton. But he was the right man for the job.

The costs of the expedition would now be the responsibility of Halley and the Royal Society, and Sir John Hoskins, the fourth President of the Royal, handsomely provided a bond of £600 for the purpose. On 19 June 1696, the Navy Board received a letter from

Halley detailing the ship's company of 15 men, 2 boys, himself, 'Mr Middleton and his servant', making 20 in all. But this is the last we hear of Middleton, and once again the ship did not sail as planned. Nothing more is heard of Middleton (perhaps the further delay was caused by him abandoning the project, or by the continuation of the war with France). On the brink of setting sail, the expedition suffered another delay, and in August 1696 the *Paramore* was laid up in wet dock.

The delay was linked to another twist in Halley's career. He became Newton's Deputy Comptroller at the Chester Mint. It is possible that the delay was caused by this appointment; but it seems more likely that Newton offered the job because he knew that Halley was at a loose end.

Throughout 1695, Halley's main scientific interest had been comets, and he had carried out a correspondence on the subject with Newton. He showed that at least some comets followed elliptical orbits, obeying the same laws as planets in their repeated movement around the Sun, while others followed parabolic orbits, so that they dived once past the Sun before departing for the depths of space, never to return. He was keenly interested in the comet he had seen (along with many other people) in 1682, and had begun to suspect that it had been seen at least twice before, in 1531 and 1607, implying that it was in an orbit around the Sun with a period of seventy-five or seventy-six years. Halley also appreciated that the orbit of a comet might not repeat exactly, because of disturbances caused by the gravitational influence of Saturn and Jupiter. As he wrote to Newton:

> I must entreat you to consider how far a Comets motion may be disturbed by the Centers of Saturn and Jupiter, particularly in its ascent from the Sun, and what difference they may cause in the time of the Revolution of a Comett in its so very Elliptick Orb.

Newton wrote back:

> I can never thank you sufficiently for this assistance & wish it in my way to serve you as much.

It is possible that the job at the Chester Mint was one way Newton had of returning the favour (it's also possible that he simply thought

Halley was the best man for the job). In the last of this flurry of letters to Newton on the subject of comets, undated but probably written in March 1696, Halley says: 'I will waite on you at your lodgings to morrow morning to discourse the other matter of serving you as your Deputy.'

So by 1696 Halley knew that at least some comets move in elliptical orbits, and he suspected that some of these, in particular the one seen in 1682, returned at regular intervals. None of this was published in the mid-1690s, although he did talk about these ideas at the meetings of the Royal in June and July 1696, and he must have continued to calculate the orbits of various comets over the next few years, since he published the details of some twenty cometary orbits in 1705. In one way, this must have been a frustrating time for Halley. He knew that Flamsteed had made accurate observations of the comet of 1682, and needed those data to check his theory; but Flamsteed was not on speaking terms with Halley. Flamsteed was, though, on good terms with Newton, and Halley hoped that Newton could get the observations from Flamsteed and pass them on to him. Flamsteed, however, hated sharing his data with anyone, as we shall see.

It is, incidentally, in a letter from Flamsteed to Newton, dated 7 February 1695, that we get a hint of why the puritanical Flamsteed had taken against Halley; he writes that Halley has 'almost ruined himself by his indiscreet behaviour' and alludes to deeds 'too foule and large for [discussion in] a letter'. But there is no record of what those deeds were.

Whatever, in 1696 Halley and Newton were firm friends, with a common interest in comets. Newton was well aware how much he owed Halley in getting the *Principia* published, and may have felt slightly bad about Gregory getting the Oxford job instead of Halley. In 1696, as we have seen, Newton was appointed Warden of the Royal Mint, and given the task of overseeing a much needed reform of the currency and recoinage. One of his first acts as Warden was to appoint Halley, in the late summer of 1696, to be Deputy Comptroller of the subsidiary Mint at Chester.

This was always going to be a temporary job to see the reform of the currency through, and obviously Newton thought he was doing Halley a favour by providing him with an income of £90 per annum, if only for a couple of years.[5] Unfortunately, Halley did not enjoy his time at the Mint in Chester, where he encountered insubordination and resistance to his attempts to eradicate dishonest practices (it was all too easy for staff to line their own pockets). He did, as ever, carry out many scientific observations during his time in Chester (including studying the local magnetic variation), and he continued to act as Clerk to the Royal, although for the first time this work suffered because of his other commitments. On 25 October 1697, he wrote to Sir Hans Sloane at the Royal:

> My heart is with you and I long to be delivered from the uneasiness I suffer here by ill company in my business, which at best is but drudgery, but as we are in perpetual feuds is intollerable.

Newton knew that Halley was unhappy in Chester, and offered to find him other posts on more than one occasion; but Halley stuck it out and stayed in his position until early in 1698, when he returned to London.

Now, there was no need to delay the voyage in the *Paramore* any longer. The Nine Years War had ended with the Peace of Ryswick in October 1697, and official enthusiasm for the expedition had increased so much that the ship was to sail as a Royal Navy vessel, under the patronage of William III, at no cost to the Royal Society. As a navy ship, the *Paramore* had guns mounted, and a navy crew. But there seems to have been no question raised about Halley's appointment, and he remained Master and Commander. The *Paramore* was rigged and ready to sail in March 1698. But before the great voyage could begin, Halley was given another job which shows how well regarded he was in nautical matters and as a representative of the Crown. Czar Peter I of Russia (1672–1725), the great modernizer (and later known as Peter the Great), was touring the countries of western Europe to study modern practices in many fields, before returning to Moscow and applying them at home. One of his interests was the navy, and he was visiting England with the express wish of studying English shipbuilding. The Admiralty gave him access to their

5. If Halley did lose heavily in the Lagos disaster, he may have needed the money at this time.

dockyards at Deptford, and provided several ships, including the *Paramore*, for him to sail on.

Peter stayed in Deptford for a time, literally getting his hands dirty (and calloused) as he studied how ships were built there. His heavy drinking, womanizing, and crude manners became the stuff of many stories, and he and his entourage wrecked John Evelyn's house, which they had borrowed from 6 February to 21 April. There are hazy anecdotal accounts that Halley dined with the Czar on more than one occasion, and joined in some of the wilder games, which included being pushed in a wheelbarrow through the ornamental hedges of the garden. When Peter left, the Exchequer was obliged to pay Evelyn £300 in compensation for all the damage (the bill included the cost of three broken wheelbarrows, and the total sum can be put into perspective when you remember that the expected cost of the twelve-month voyage of the *Paramore* was no more than £600). Clearly, if Halley was a friend of Czar Peter, it is easy to see why he was no friend of the prim and proper Flamsteed.

The trips with Czar Peter on board had helped to form part of the sea trials of the *Paramore*, and she had been shown to need further work before she would be fit for a long voyage. After further trials, the necessary orders were issued on 3 July. With the work completed, on 15 October 1698 Halley was issued with instructions for a year-long voyage aimed at measuring magnetic variations and thereby providing a tool for navigation (the instructions came as no surprise to Halley; he had drafted them, and given them to the Admiralty, which then passed them back to him as orders).[6]

You are to make the best of your way to the Southward of the Equator, and there to observe on the East Coast of South America, and the West Coast of Africa, the variations of the Compasse, with all the accuracy you can, as also the true Scituation both in Longitude and Latitude of the Ports where you arrive.

You are likewise to make the like observations at as many of the Islands in

6. Somewhat confusingly, on 19 August 1698 Halley had been issued with a second commission as Master and Commander of the *Paramore*. This was the same as the commission issued in 1696. It may have been felt that the original one had lapsed in some way, or it may reflect the fact that from now on the voyage is clearly an Admiralty affair, and no longer a privately funded expedition.

the Seas between the aforesaid coasts as you can (without too much deviation) bring into your course; and if the Season of the Yeare permit, you are to stand soe far into the South, till you discover the Coast of the Terra Incognita, supposed to lye between Magelan's Streights and the Cape of Good Hope, which Coast you are carefully to lay down in its true position.

In your return you are to visit the English West India Plantations, or as many of them as conveniently you may, and in them to make such observations as may contribute to lay them downe truely in their Geographicall Scituations.

The ship set off on its voyage on 20 October 1698. Halley, now well and truly a sea captain, was a few days short of his 42nd birthday, and about to have his authority tested in no uncertain fashion.

You might think that as a landsman entrusted with a King's ship, Halley would have been given a First Lieutenant he could rely on, a hand-picked trusty seafarer who could look after the actual running of the ship. Not a bit of it. Somehow, through indifference or carelessness, or both, the Admiralty managed to saddle Halley with one Edward Harrison, a fine seaman and competent officer with eight years' service, but somebody who just happened to have a grudge against the Royal Society in general and Halley in particular.[7] In 1694, Harrison had submitted a paper on the longitude problem to the Royal, and they had rejected it. In 1696, Harrison published his ideas in a little book, and the Admiralty appointed a committee of experts, including Halley, to look into it. The committee decided that Harrison's ideas were useless. So in addition to the natural unhappiness any career naval officer might have felt at being ordered to serve under a landsman, Harrison had every incentive to show Halley up as an unsuitable commander and an incompetent navigator – he may even have engineered his appointment for that purpose.

In spite of the difficulties caused by Harrison, which were initially carefully kept short of open disobedience to orders, the *Paramore* reached Brazil and the West Indies and carried out some of Halley's planned scientific work. But in April and May 1699, in the West Indies, matters came to such a head that Halley decided to cut the voyage short and head home. After he reached London, Halley

7. He was, though, no relation to the John Harrison of longitude fame.

explained what had happened in a letter to Josiah Burchett, who had succeeded Pepys as Secretary to the Admiralty:

On the fifth of [June 1699] [Harrison] was pleased so grosly to affront me as to tell me before my Officers and Seamen on Deck, and afterwards owned it under his hand, that I was not only uncapable to take charge of the Pink, but even of a Longboat; upon which I desired him to keep his Cabbin for that night, and for the future I would take the charge of the Shipp my self, to shew him his mistake; and accordingly I have watched in his steed ever since, and brought the Shipp well home.

To Harrison's undoubted chagrin, Halley navigated the ship home perfectly (proving that if you want a good navigator you should hire an astronomer), whereupon Harrison was court-martialled, but escaped with a reprimand, since there was no proof that he had disobeyed a direct order. He may seem to have got off lightly, but discipline in the Royal Navy at the end of the seventeenth century wasn't quite so strict as it would later become, and such criticisms of superior officers were not particularly uncommon, not least because of the system of appointing gentlemen over the heads of tarpaulins. A century later, Harrison wouldn't have got off so lightly – but a century later, a landsman like Halley, whatever his skills as a navigator, would not have been given command of a King's ship. If nothing else, though, the whole episode sheds some light on Halley's powerful personality and determination; not a man to get in the way of.

Halley took advantage of the brief time he spent back in England to resign his post as Clerk to the Royal Society, where deputies had been standing in for him during his absence, clearing the way for his re-election as a Fellow, and set off again in the *Paramore*, now in unquestioned command, on 16 September 1699 (the first voyage had ended only on 28 June). He carried out his magnetic observations all the way down to latitude 52° South (almost level with the tip of South America), encountering huge icebergs. He also made meteorological observations, and carried out important charting work. The ship returned to Plymouth, in good order, on 27 August 1700.[8]

8. If you want to know more about Halley's voyages, see *The Three Voyages of Edmond Halley*, edited by Norman Thrower.

His return was a triumph. As well as reporting to the Admiralty and publishing charts of his voyages, Halley presented papers on his discoveries made on the two voyages to the Royal Society (where he was re-elected as a Fellow on 30 November 1700), and was almost ready to take up his career as a scientist again, but without the time-consuming duties of the Clerkship. Almost, but not quite. In 1701, he asked the Admiralty if they would lend him a ship to carry out a study of the tides in the English Channel. They responded with alacrity, giving him command of the *Paramore* again, and he set out on 14 June that year, completing his work and paying off the crew of the pink on 16 October (no naval officer except Halley ever commanded the *Paramore* on a lengthy voyage; in 1706 she was sold for £122). But it seems there was more to this last voyage than meets the eye. As well as the genuine scientific work, Halley, under orders from the Admiralty, carried out some clandestine surveying of French ports and harbour approaches, for use in case of war.

The powers that be were obviously pleased by the discreet way Halley carried out this work, and in 1702, after Queen Anne succeeded William III, Halley was sent on another expedition for the Crown.

Once again, Europe was at war. This time, the argument was about who should be King of Spain. Charles II of Spain had died in 1700, naming the 17-year-old grandson of Louis XIV, who became Philip V of Spain, as his heir. This concentration of power in the hands of the Bourbons provoked action by the Holy Roman Empire, the Dutch, and the English, triggering the War of the Spanish Succession in 1702. It was in this war that a combined Anglo-Dutch fleet captured Gibraltar from Spain, and became a powerful force in the Mediterranean, where Spain ruled over a large part of Italy, as well as the Iberian peninsula. But before the capture of Gibraltar, England was looking for other bases in the region. The Holy Roman Empire, England's ally on this occasion, owned ports in the northern Adriatic, and Halley was chosen as an envoy to Vienna, to survey these ports and investigate the feasibility of using them as bases for the allied fleets in the campaigns against Spain and France. This involved diplomacy (where his status as a Colonel came in useful), hydrographic surveys of the kind he was familiar with, and overseeing the practical aspects

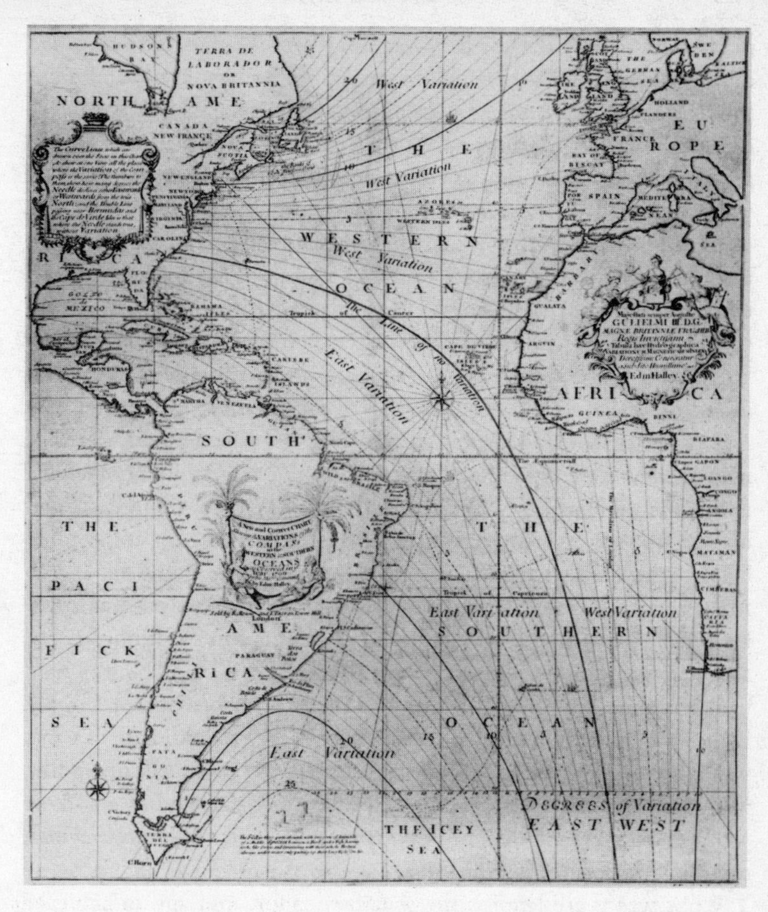


*Halley's second voyage around the Atlantic*

of construction of the fortifications that were planned. He travelled overland to Vienna, where he arrived on 10 January 1703, and on to the Adriatic ports, returning to London before the end of April. He must have made a good impression, since the Emperor, Leopold I, presented him with a valuable diamond ring on the return journey.

He returned on a second mission to check that the work was being carried out properly, leaving London on 22 June. On 16 June, George Stepney, the representative of Queen Anne in Vienna, had written to Sir Charles Hedges, Secretary of State in London, that 'it will be an ease to me to have Cap^t. Halley in these parts to look after the Fortifications and Provisions which are things I do not understand'.

Both trips involved extensive travel across Europe and meetings with many eminent people. This time, on the way to Vienna, Halley had dinner in Hanover with the future King George I of England, and his son (the future George II). It would have been natural to keep the heir to the English throne informed of what was going on in the Adriatic ports. Halley's work also seems to have been used as cover for a little spying; he finally returned to England in November 1703, and a letter dated 14 January 1704 from the Earl of Nottingham instructs the Chancellor of the Exchequer to pay the sum of £36 for Halley's expenses 'out of the secret service'. All round, Halley's work was a great success. The fleets never set up home in the Adriatic ports, however, because the capture of Gibraltar in 1704, and also of Port Mahon, on Minorca, in 1706, gave them even better bases from which to attack the French and Spanish.

What was there left for the scientist, sailor, and spy to achieve? Only his dearest wish – an academic post. On 28 October 1703, just before Halley returned from his second diplomatic mission across Europe, John Wallis, who had been the Savilian Professor of Geometry at Oxford University since 1649, and was one of the first Fellows of the Royal Society, died.[9] Who else but Halley could replace him? He still had his opponents, notably Flamsteed, who wrote in a letter to Abraham Sharp dated 18 December 1703 that 'Dr Wallis is dead –

9. This was the same year that Robert Hooke and Samuel Pepys died. By the end of 1703, the foundation of the Royal Society was becoming history, not a living memory.

Mr Halley expects his place – who now talks, swears, and drinks brandy like a sea captain'. Well, after all, Halley *was* now a sea captain! But he was at the height of his fame, with far more supporters than opponents. On 20 November, shortly before Halley's return to England, the Earl of Nottingham, Secretary of State alongside Hedges, wrote to Viscount Hatton (the head of the family to which the Earl belonged) that:

> Everyone who has a vote in the election[10] thinks of him . . . I have seen his zeal in the publick service (on which he is now abroad) the which added to his extraordinary skill above all his competitors obliged me to be very forward in promoting him to this place.

At the end of November 1703, Halley was elected to the Council of the Royal Society, the same day that Newton became its President; and in January 1704, at the age of 47, he was duly appointed to the professorship in Oxford. He was formally admitted to the post on 7 March, and gave his inaugural lecture on 24 May. To his delight, around Oxford he was generally referred to as 'Captain Halley', at least until 1710, when he received a degree of Doctor of Civil Law from Oxford, and became 'Doctor Halley' to all – at that time this seems to have been regarded as a more important title than 'Professor'. There are good reasons why Halley should have been particularly proud of his doctorate. The degree was, in fact, an honorary one, a reward for Halley's many years of public service, and unlike the doctoral candidates who achieved that status through their academic work, Halley paid no fees for the honour.

Once he became a respectable, and respected, academic professor, Halley's life became relatively uninteresting. There would be no more deep-sea diving, naval voyages, or spying expeditions. His book *A Synopsis of the Astronomy of Comets*, largely based on the work he had carried out ten years earlier, was published in 1705, and included the famous prediction of the return of the comet of 1682, which Halley suggested would occur 'about the year 1758' long after the time Halley himself expected to be dead. During his time in Oxford, Halley did study (among other things) eclipses, meteors, and the

10. Nottingham was one of the nine electors for the post.

aurora, as well as doing important work in mathematics; but his most important contribution (apart from his theory of comets) was in stellar astronomy.

This harked back to his first major piece of work, the map of the southern skies, and also passed through an unfortunate episode involving Flamsteed's observations.

The Royal Greenwich Observatory had been set up expressly so that Flamsteed could prepare more accurate astronomical tables (important in navigation), but he was extremely reluctant to publish, and hugged his results to himself. He claimed that although he was paid by the Crown, this was only a token sum and that most of the cost of the instruments at the Observatory (the RGO) and the observing itself had come out of his own pocket, so he was entitled to keep the observations and publish only when he was good and ready, and not before (in Flamsteed's defence, this was partly because he was a perfectionist who wanted to dot every *i* and cross every *t* before showing his catalogue to the world). In 1704, Newton, as President of the Royal Society, had persuaded Flamsteed to hand over some of his data, and over the next few years printing of a new star catalogue began, but ground to a halt in the face of Flamsteed's continuing prevarication. Flamsteed had become embittered and saw every hand against him. He had long ago fallen out with Halley, partly because Halley had been a good friend of Hooke, whom Flamsteed had loathed, and he even took against David Gregory, an inoffensive man who, as Savilian Professor of Astronomy in Oxford, ought to have been Flamsteed's closest professional colleague.

By 1710, everybody had had enough, and Queen Anne issued a Royal Warrant which appointed Newton and such other Fellows of the Royal Society as he chose to act as a Board of Visitors to the RGO, with the authority to demand all of Flamsteed's results so far, and to have a fair copy of his continuing annual observations within six months of the end of each year. Even Flamsteed couldn't argue with a Royal Warrant, and Newton gave Halley the task of knocking the raw material from Flamsteed, including his old data from 1704, into shape for publication.

Halley had to add a lot of data of his own to fill the gaps in the material supplied by Flamsteed, and this effort renewed his interest in making accurate observations of the positions of the stars. He made every effort to get Flamsteed to cooperate in the project, including sending him the proofs for checking, but without success. The Flamsteed saga raged on – the version of the catalogue edited by Halley appeared in 1712 under the title *Historia Coelestis*, and eventually Flamsteed (now in his late sixties) and his wife managed to buy up most of the copies that were printed and burn them. The incident did, however, encourage him to put his data into a form he was happy with; although the resulting catalogue was not published until 1725, six years after Flamsteed's death, it gave nearly three thousand star positions to an accuracy of 10 arc seconds, a great improvement over earlier catalogues.

Meanwhile, starting around the time he began work on Flamsteed's data, Halley made a careful investigation of the data in a star catalogue included in the work of Ptolemy, based on material originally gathered by Hipparchus and dating from the second century BC. He found that most of the star positions given in the old catalogue closely matched observations made in Halley's time, but a few were considerably different from the positions seen in the eighteenth-century sky. Because most of the positions did agree with his contemporary data, Halley realized that the few 'errors' were not really observational mistakes, and that these stars had actually moved across the sky since the time of Hipparchus. The star Arcturus, for example, appeared in the early eighteenth century twice the width of the full Moon (more than a degree of arc) away from the position recorded by Ptolemy. This was the first direct evidence that stars are scattered across space and move independently, not being attached to a single crystal sphere around the Earth.

The most dramatic piece of work that Halley was associated with at this time concerned his studies of the total eclipse of the Sun seen in 1715 (the occasion when Kitty Barton made such an impression on Rémond de Montmort). This was the first such eclipse to be visible from London since 1104, and had been predicted long in advance. Halley was responsible for organizing observers across England to report exactly what they saw and to send their observations to the Royal Society. As a result, the path of the eclipse was recorded with unprecedented accuracy.

This proved invaluable to astronomers in the late twentieth century, when they became interested in the possibility that the size of the Sun may have changed slightly over the centuries. Because we now know the distances to the Sun and Moon (and the size of the Moon) with great accuracy, measurements of the width of the path of totality during a solar eclipse can be used to calculate geometrically the size of the Sun itself. Halley's data were good enough to show, when compared with observations of more recent eclipses, that the Sun does indeed 'breathe' in and out by a tiny amount on a timescale of centuries. It has been suggested that these changes may be linked to the period of severe cold experienced during Halley's lifetime. But this wasn't the first occasion that Halley's data had proved invaluable long after his death.

On 31 December 1719 Flamsteed died, and on 9 February 1720 Halley was appointed as the second Astronomer Royal, a post he held alongside his professorship in Oxford, by King George, who had ascended to the throne in 1714. Now 63, the first thing he had to do was refurbish the RGO, which had been stripped of its instruments and furniture by Flamsteed's widow, who claimed them as part of his estate (Halley received a grant of £500 for the refurbishment; this included installing one of the old telescopes made by Hooke). Halley might have been expected to settle down to a quiet old age as the grand old man of British science (especially after Newton died, in 1727); but not a bit of it. As well as continuing observations of phenomena such as nebulae (fuzzy patches of light on the sky, some now known to be clusters of stars) and variable stars, in 1722 he began work on a project to map the changing position of the Moon against the background stars over an entire eighteen-year cycle, the saronic cycle, the period with which this motion repeats.

Halley had begun studying the Moon when he was still a schoolboy (but one, remember, equipped with professional-quality equipment). Every astronomer of his generation studied the Moon, either investigating its motion against the background stars or mapping its surface features, or both. Halley had always been especially interested in the way the Moon moved, which long observations (even before the days of astronomical telescopes) had shown followed this regular

*The solar eclipse of 1715*

eighteen-year cycle. Before Newton and the *Principia*, these studies were entirely empirical, and there was no scientific theory of why the Moon should follow its observed orbit. After the *Principia*, lunar studies remained at the forefront of astronomy, partly because they provided a test bed for Newton's theory of gravity. But there was another, practical reason to study the orbit of the Moon, even if Newton's theory had not existed. And it was close to Halley's heart as a sailor.

The aim was to establish a technique for measuring position at sea from observations of the Moon made at Greenwich. If the orbit of the Moon was known precisely for its entire saronic cycle, then an observer anywhere on Earth could tell the exact time at Greenwich (or some other chosen baseline) by measuring the position of the Moon against the stars – not just the day or week, but the exact hour, or even (in principle) the exact minute. Comparing Greenwich time with local noon would tell a navigator how far east or west of Greenwich the ship was. The whole procedure was horribly complicated, and it's a measure of how important the problem of navigation at sea was in the seventeenth century that so much effort was devoted, by Halley and others, to trying to make it work. It never did become a practical method of finding longitude at sea, but Halley wasn't to know that when he set out in 1722 on the task of studying the orbit of the Moon for as much of a saronic cycle as he could.

Halley did not expect to live to complete the task, but he did. Although the observations were never used in navigation, this does not diminish the skill with which they were made by the elderly astronomer. The key to navigation at sea turned out to be the development of accurate portable chronometers by John Harrison (the namesake of Halley's former lieutenant), who visited Halley at Greenwich in 1728 to discuss the problem with him. Halley was a member of the Board of Longitude, commissioned by Queen Anne in 1714 to be the judges of a competition to find a method for determining longitude at sea. A prize of £10,000 was offered for a method accurate to within 60 miles on a voyage to the West Indies and back, with £15,000 if the error were less than 40 miles and £20,000 (equivalent to several million pounds in today's money) if it were accurate to 30 miles (equivalent to a timing error of about 12 seconds on the round trip). The tale of how John Harrison eventually received the prize has been told by Dava Sobel, in her book *Longitude*.[11]

In spite of his age, Halley remained active in astronomy, and published scientific papers, although nothing to rank with his earlier work, for another seventeen years after he became Astronomer Royal. He was also active in the Royal Society, where he was elected as Vice-President in 1731, and regularly attended meetings even in his seventies. Although his wife died on 30 January 1736 (the exact cause is not known) and Halley suffered a stroke at about the same time, leaving him with slight paralysis in his right hand, he remained active as an observer almost until his death, on 14 January 1742, a few weeks after his 85th birthday. This came at the end of a short illness and gradual loss of strength. We are told that he died while sitting in a comfortable chair, shortly after enjoying a glass of wine.

Halley's later years had been quite comfortable financially, even though (as Flamsteed had rightly pointed out) the income from the post of Astronomer Royal was small. He still had his Oxford post (although towards the end of his life this was in effect a sinecure), and benefited in 1729 from a visit to the RGO by Queen Caroline (the wife of George II, whom Halley had met in Hanover, and who had succeeded his father in 1727). The Queen was intrigued to learn about Halley's former career as the Master and Commander of a King's ship, and that he had served in that capacity for more than three years. At the time, three years in command qualified the officer for a pension of half the pay appropriate to the rank of Post Captain. The Queen ensured that Halley received his naval pension until his death. He was buried, alongside Mary, in the churchyard at Lee, close to the Observatory; the original church was demolished and replaced in 1840, but the original tombstone was moved to the Observatory in 1854, and set into a wall there.

Halley's fame, and his position in the popular imagination, was ensured nearly seventeen years after his death, when the comet now known by his name returned as predicted. Halley himself

11. Fourth Estate, London, 1996.

had been aware that the date he gave for the return, 1758, was only approximate, because the orbit of the comet is constantly being disturbed by the gravitational influence of the planets, notably the giant Jupiter. Its return was actually first seen by a German farmer and amateur astronomer, Georg Palitzsch, on Christmas Day, 1758; but this was not widely known at the time, and the first detection made by a professional astronomer was the work of Charles Messier, at the Paris Observatory (the French counterpart to the Royal Greenwich Observatory), on 21 January 1759. Astronomers nowadays give the date of the return as 1759, not out of any disrespect to Palitzsch but counting from its closest approach to the Sun (perihelion), on 13 April that year. The time of perihelion is always a well-determined date, for the passage of any comet around the Sun, whereas the time of first sighting owes much to luck.

Halley's prediction of the return of the comet had been widely recognized, both in Britain and in continental Europe, as the definitive test of both Newton's theory of gravity and Newtonian mechanics. By the second half of the eighteenth century, the scientific principles spelled out so long before by William Gilbert and Francis Bacon had become well established, and it was understood that a scientific theory was only a good one if it made accurate predictions about the outcome of experiments or the behaviour of the world. The prediction had to be something new – it would be no good saying that you predict the Sun will rise in the east tomorrow, and then claiming that this is a good scientific theory when the Sun does rise in the east as predicted. The Sun has, as far as we know, always risen in the east, so there is nothing to the prediction. Even *explaining* long-established facts, like the elliptical shape of the orbit of Mars, was not enough to establish a theory; *predicting*, accurately, something that nobody had seen before, the date on which a comet would appear in the sky, was quite another matter. And somehow this particular prediction seemed even more impressive since both Newton and Halley were dead; the scientists who did the work that made the prediction possible had gone, but the scientific truth remained for all time.

The French astronomer Joseph de LaLande, who later became Director of the observatory in Paris, wrote in 1759 that:

The universe beholds this year the most satisfactory phenomenon ever presented to us by astronomy; an event which unique until this day changes our doubts to certainty and our hypotheses to demonstration.

It was, indeed, the moment when science came of age, and the scientific revolution was complete.

But Halley's posthumous contributions to science weren't finished with the return of 'his' comet, and de LaLande wasn't finished with Halley's predictions. In 1761, the transit of Venus across the face of the Sun was observed, using the techniques Halley had spelled out back in 1716, from a total of 62 observing stations. A similar number of observing stations monitored the transit of 1769; one crucial set of observations of the second transit was made from the Pacific island of Tahiti, on the famous voyage of Captain James Cook, a voyage carried out largely at the urging of the Royal Society. The combined results of all these measurements were used to work out the distance from the Earth to the Sun. LaLande, who had been born in 1732 and lived until 1807, was the person chosen to collect and collate all the data. When the calculations were complete, they gave a value for the distance to the Sun of 153 million kilometres, very close to the modern value of 149.6 million kilometres. In this way, Halley made his last major contribution to astronomy twenty-seven years after he had died.

It is strange that somebody who is literally a household name, through his association with comets in the public eye, is also one of the most underrated scientists of all time. Although highly regarded in his own lifetime, for two hundred years following his death Halley's achievements, like those of Hooke, have largely been overshadowed by those of Newton, when in any other generation he would have stood out as the greatest astronomer – perhaps the greatest scientist – of his time. The 'Newtonian Revolution' in science was actually the work of three men whose careers overlapped in the second half of the seventeenth century and were all intimately associated with the Royal Society – Isaac Newton himself, Robert Hooke, and Edmond Halley.

Halley, for all his achievements, clearly ranks third in terms of his contribution to science. The traditional view has been that equally clearly Newton ranks first. But if Newton had not existed, someone else (or several other people) would have taken the steps he took within the next few years. After all, Leibnitz *did* invent the calculus, and Huygens (for example) was quite capable of working out the inverse-square law. Without Hooke, though, the Royal Society would probably have died an early death, or become merely a gentlemen's talking shop. And without the Royal, the revolution in science would have occurred at another time (perhaps only a little later) and another place (perhaps in Paris), and in a different way. It was Hooke who put into action the experimental scientific method that was central to the philosophy of the Royal, a method they believed they owed to Bacon but which really had its roots in the work of Gilbert and Harvey. It was Newton who appreciated the literally universal application of the scientific method. And it was Halley who proved that Newton was right. The revolution in science was, indeed, not the work of one man, but of a Fellowship.

# *Sources and Further Reading*

Angus Armitage, *Edmond Halley*, Nelson, London, 1966.

John Aubrey, *Brief Lives*, ed. Andrew Clark, volumes I and II, Clarendon Press, Oxford, 1898; see also the single-volume abridgement, ed. Oliver Lawson Dick, Secker & Warburg, London, 1949.

Jim Bennett, Michael Cooper, Michael Hunter, and Lisa Jardine, *London's Leonardo*, Oxford University Press, 2003.

Thomas Birch, *The history of the Royal Society of London*, Royal Society, London, 1757.

F. F. Certore, *Robert Hooke's Contributions to Mechanics*, Martinus Nashoft, The Hague, n.d.

Alan Cook, *Edmond Halley*, Oxford University Press, 1998.

John Craig, *Newton at the Mint*, Cambridge University Press, 1946.

J. G. Crowther, *Founders of British Science*, Cresset Press, London, 1960.

J. G. Crowther, *Francis Bacon*, Cresset Press, London, 1960.

William Dampier, *A History of Science*, 3rd edition, Cambridge University Press, 1942.

William Derham, ed., *Philosophical Experiments and Observations of Robert Hooke*, London, 1726.

René Descartes, *Discourse on Method and the Meditations*, trans. F. E. Sutcliffe, Penguin, London, 1968.

John Donne, *Donne's Sermons*, ed. George Potter and Evelyn Simpson, University of California Press, Berkeley, 1951.

E. T. Drake, *Restless Genius: Robert Hooke and his Earthly Thoughts*, Oxford University Press, 1996.

Stillman Drake, *Galileo at Work*, Dover, New York, 1978.

Stillman Drake, *Galileo*, Oxford University Press, 1980.

Margaret 'Espinasse, *Robert Hooke*, Heinemann, London, 1956.

John Evelyn, *Diary*, ed. E. S. de Beer, 6 volumes, Clarendon Press, Oxford, 1955.

J. J. Fahie, *Galileo: his life and work*, Murray, London, 1903.
Benjamin Farrington, *Francis Bacon*, Lawrence & Wishart, London, 1951.
Benjamin Farrington, *The Philosophy of Francis Bacon*, Liverpool University Press, 1970.
J. F. Fulton, *Sir Kenelm Digby. Writer, bibliophile and protagonist of William Harvey*, Oliver, New York, 1937.
Galileo Galilei, *Galileo on the World Systems*, abridged and trans. from the *Dialogue* by M. A. Finocchiaro, University of California Press, 1997.
William Gilbert, *On the Loadstone and Magnetic Bodies, and on The Great Magnet of the Earth*, trans. from the 1600 edition of *De Magnete* by P. Fleury Mottelay, Bernard Quaritch, London, 1893.
William Gilbert, *De Magnete*, trans. P. Fleury Mottelay, Dover, New York, 1958; reprint of edition first published in 1893.
John Gribbin, *Science: A History*, Allen Lane, London, 2002.
R. T. Gunther, *Early Science in Oxford*, Oxford University Press, 1930.
Marie Boas Hall, *Robert Boyle and Seventeenth-Century Chemistry*, Cambridge University Press, 1958.
Rupert Hall, *The Revolution in Science 1500–1750*, Longman, London, 1983.
Rupert Hall, *Isaac Newton: Adventurer in thought*, Blackwell, Oxford, 1992.
R. Harré, *Early Seventeenth Century scientists*, Pergamon, Oxford, 1965.
R. Harré, *The Method of Science*, Wykeham, London, 1970.
Harold Hartley, ed., *The Royal Society – Its Origins and Founders*, Royal Society, 1960.
William Harvey, *Movement of the Heart and Blood*, trans. K. J. Franklin, Oxford University Press, 1957.
William Harvey, *De motu locali animalium 1627*, ed. and trans. Gweneth Whitteridge, Cambridge University Press, 1959.
Robert Hooke, *Micrographia*, Royal Society, London, 1665; available in facsimile edition (of original 1665 edition), Dover, 1961.
Robert Hooke, *The Posthumous Works of Robert Hooke*, ed. Richard Waller, Royal Society, London, 1705.
Robert Hooke, *Lectures and Discourses of Earthquakes and Subterraneous Eruptions*, Arno Press, New York, 1978; reprinted from *Posthumous Works* (1705) and includes Waller's *Life*.
Robert Hooke, *The Diary of Robert Hooke*, ed. Henry Robinson and Walter Adams, Taylor & Francis, London, 1935.
William Huggins, *The Royal Society*, Methuen, London, 1906.
Lynette Hunter and Sarah Hutton, eds., *Women, Science and Medicine 1500–1700*, Sutton, Stroud, 1997.

Michael Hunter, *The Royal Society and its Fellows 1660–1700*, British Society for the History of Science, Chalfont St Giles, 1982; revised edition published 1994.

M. Hunter and S. Schaffer, eds., *Robert Hooke: New Studies*, Boydell Press, Woodbridge, 1989.

Robert Hutchins, ed., *Gilbert, Galileo, Harvey*, Encyclopedia Britannica, Chicago, 1952; reprints in English of key works of each of the three pioneers of science.

Stephen Inwood, *The Man Who Knew Too Much*, Macmillan, London, 2002.

Lisa Jardine, *On a Grander Scale: The Outstanding Career of Sir Christopher Wren*, HarperCollins, London, 2002.

Lisa Jardine, *The Curious Life of Robert Hooke*, HarperCollins, London, 2003.

Lisa Jardine and Alan Stewart, *Hostage to Fortune: The Troubled Life of Francis Bacon*, Gollancz, London, 1998.

Kenneth Keele, *William Harvey the Man, the Physician, and the Scientist*, Nelson, London, 1965.

Suzanne Kelly, *The De Mundo of William Gilbert*, Hertzberger, Amsterdam, 1965.

Geoffrey Keynes, *A Bibliography of Dr Robert Hooke*, Clarendon Press, Oxford, 1960.

Geoffrey Keynes, *The Life of William Harvey*, Clarendon Press, Oxford, 1966.

Alexandre Koyré, *Newtonian Studies*, University of Chicago Press, 1965.

Henry Lyons, *The History of the Royal Society 1660–1940*, Cambridge University Press, 1944.

Thomas Macaulay, *Critical and Historical Essays*, Longman Green, London, 1877.

E. F. MacPike, ed., *Correspondence and Papers of Edmond Halley*, Clarendon Press, Oxford, 1932.

Frank Manuel, *Portrait of Isaac Newton*, Harvard University Press, 1968.

Louis More, *Isaac Newton*, Scribner's, New York, 1934.

Isaac Newton, *Newton's Philosophy of Nature*, selections from his writings, ed. H. S. Thayer, Hafner, New York, 1953.

Markku Peltonen, ed., *The Cambridge Companion to Bacon*, Cambridge University Press, 1996.

Samuel Pepys, *The Shorter Pepys*, selected and ed. Robert Latham, Penguin, London, 1987.

Roger Pilkington, *Robert Boyle: father of chemistry*, Murray, London, 1959.

D'Arcy Power, *William Harvey*, Unwin, London, 1897.

Lawrence Principe, *The Aspiring Adept*, Princeton University Press, 1998.

Margery Purver, *The Royal Society: Concept and Creation*, Routledge, London, 1967.

James Reston, *Galileo*, Cassell, London, 1994.

H. W. Robinson and W. Adams, eds., *Diary of Robert Hooke*, London, 1935.

Duane Roller, *The De Magnete of William Gilbert*, Hertzberger, Amsterdam, 1959.

Colin Ronan, *Edmond Halley*, Macdonald, London, 1970.

Leona Rosenberg, *The Library of Robert Hooke*, Madon Press, Santa Monica, 1989.

Steven Shapin, *The Scientific Revolution*, University of Chicago Press, 1966.

James Spedding, *The Letters and the Life of Francis Bacon*, 7 volumes, Longman, London, 1861–74.

James Spedding, Robert Ellis, and Douglas Heath, eds., *The Works of Francis Bacon*, 7 volumes, Longman, London, 1857–9.

Thomas Sprat, *History of the Royal Society*, ed. Jackson Cope and Harold Jones, Routledge, London, 1959; facsimile of original 1667 edition.

Norman Thrower, ed., *The Three Voyages of Edmond Halley*, Hakluyt Society, London, 1980.

H. W. Turnbull, ed., *The Correspondence of Isaac Newton*, 7 volumes, Royal Society, London, 1959.

Conrad von Uffenbach, *London in 1710, from the Travels of Zacharias Conrad von Uffenbach*, trans. and ed. W.H. Quarrell and Margaret Mare, Faber & Faber, London, 1934.

Brian Vickers, ed., *English Science, Bacon to Newton*, Cambridge University Press, 1987.

Dorothea Waley, *Giordano Bruno*, Singer, New York, 1950.

Richard Westfall, *Never at Rest: a biography of Isaac Newton*, Cambridge University Press, 1980; a shorter version of this book was published under the title *The Life of Isaac Newton*, by CUP, in 1993.

Peter Whitfield, *Landmarks in Western Science*, British Library, London, 1999.

Gweneth Whitteridge, ed. and trans., *The Anatomical Lectures of William Harvey*, Macdonald, London, 1964.

Gweneth Whitteridge, *William Harvey and the Circulation of the Blood*, Macdonald, London, 1971.

Robert Willis, trans., *The Works of William Harvey, M.D.*, Sydenham Society, London, 1847.

# Index

# PENGUIN SCIENCE

**SCIENCE: A HISTORY** JOHN GRIBBIN

'Brilliant … So many scientists here come alive!' Roy Porter

Filled with pioneers, visionaries, eccentrics and madmen – from Galileo to Marie Curie to Feynman and beyond, this book is an enthralling story of the men and women who changed the way we see the world.

'Essential reading' John Cornwell, *Sunday Times*, Books of the Year

'A magnificent history… enormously entertaining' *Daily Telegraph*

'We experience his subjects' triumphs and failures as if we knew them personally … I found myself whizzing through the pages' *Sunday Telegraph*

'Gripping and entertaining … Wonderfully and pleasurably accessible' *Independent on Sunday*

**STARDUST** JOHN GRIBBIN

Every one of us is made of stardust. Everything we see, touch, breathe and smell, is the by-product of stars as they live and then die in spectacular explosions, scattering material across the universe which is recycled to become part of us. Taking us on an enthralling journey, John Gribbin shows us the scientific breakthroughs in the quest for our origins among the stars.

'[An] incredible story … Gribbin takes us through the life history of the universe … gives a sense of the almost unbelievable coincidence of physical laws and circumstances that resulted in your being able to read these words today' *Literary Review*

'Gribbin is an expert science writer and this book is as good as any he has written. Beautifully conceived and highly readable' *Independent*

'The book's pace never slackens, and the quest for clues doesn't falter … Gribbin skilfully and engagingly traces the historical sequence … rather like Sherlock Holmes reading clues' David Hughes, *New Scientist*